ROCKS AND RESOURCES

McGRAW-HILL SCIENCE
MACMILLAN/McGRAW-HILL EDITION

ROCKS AND RESOURCES

RICHARD MOYER ■ LUCY DANIEL ■ JAY HACKETT
PRENTICE BAPTISTE ■ PAMELA STRYKER ■ JOANNE VASQUEZ

NATIONAL GEOGRAPHIC SOCIETY

McGraw-Hill School Division

New York   Farmington

Program Authors

Dr. Lucy H. Daniel
Teacher, Consultant
Rutherford County Schools,
North Carolina

Dr. Jay Hackett
Emeritus Professor of Earth
Sciences
University of Northern
Colorado

Dr. Richard H. Moyer
Professor of Science
Education
University of Michigan-
Dearborn

Dr. H. Prentice Baptiste
Professor of Curriculum and
Instruction
New Mexico State
University

Pamela Stryker, M.Ed.
Elementary Educator and
Science Consultant
Eanes Independent School
District
Austin, Texas

JoAnne Vasquez, M.Ed.
Elementary Science
Education Specialist
Mesa Public Schools,
Arizona
NSTA President 1996–1997

NATIONAL GEOGRAPHIC SOCIETY
Washington, D.C.

Contributing Authors
Dr. Thomas Custer
Dr. James Flood
Dr. Diane Lapp
Doug Llewellyn
Dorothy Reid
Dr. Donald M. Silver

Consultants
Dr. Danny J. Ballard
Dr. Carol Baskin
Dr. Bonnie Buratti
Dr. Suellen Cabe
Dr. Shawn Carlson
Dr. Thomas A. Davies
Dr. Marie DiBerardino
Dr. R. E. Duhrkopf
Dr. Ed Geary
Dr. Susan C. Giarratano-Russell
Dr. Karen Kwitter
Dr. Donna Lloyd-Kolkin
Ericka Lochner, RN
Donna Harrell Lubcker
Dr. Dennis L. Nelson
Dr. Fred S. Sack
Dr. Martin VanDyke
Dr. E. Peter Volpe
Dr. Josephine Davis Wallace
Dr. Joe Yelderman

Invitation to Science, *World of Science*, and *FUNtastic Facts* features found in this textbook were designed and developed by the National Geographic Society's Education Division.
Copyright © 2000 National Geographic Society
The name "National Geographic Society" and the Yellow Border Rectangle are trademarks of the Society, and their use, without prior written permission, is strictly prohibited.

Cover Photo: ZEFA/Stock Imagery, Inc.

McGraw-Hill School Division
A Division of The McGraw-Hill Companies

Copyright © 2000 McGraw-Hill School Division,
a Division of the Educational and Professional
Publishing Group of The McGraw-Hill Companies, Inc.

All rights reserved. No part of this book may be reproduced or transmitted in any form or by any means, electronic or mechanical, including photocopying, recording, or by any information storage and retrieval system, without permission in writing from the publisher.

McGraw-Hill School Division
Two Penn Plaza
New York, New York 10121

Printed in the United States of America

ISBN 0-02-278212-5 / 3

1 2 3 4 5 6 7 8 9 058/046 05 04 03 02 01 00 99

CONTENTS

UNIT 5 — ROCKS AND RESOURCES

CHAPTER 9 • THE CHANGING EARTH 257

TOPIC 1: LOOKING UNDER YOUR FEET 258
- **EXPLORE ACTIVITY** Investigate How Rocks Are Alike and Different 259
- **QUICK LAB** Mineral Scratch Test 261
- **NATIONAL GEOGRAPHIC** WORLD OF SCIENCE Stone Symbols 266

TOPIC 2: SLOW CHANGES 268
- **EXPLORE ACTIVITY** Investigate How Rocks Change 269
- **QUICK LAB** Changing Chalk 271
- **SKILL BUILDER** Forming a Hypothesis: Which Materials Settle First? 273
- **SCIENCE MAGAZINE** Erasing Erosion 276

TOPIC 3: FAST CHANGES 278
- **DESIGN YOUR OWN EXPERIMENT** How Can Land Change Quickly? 279
- **QUICK LAB** Weather Adds Up 281
- **SCIENCE MAGAZINE** Predicting Hurricanes 286

CHAPTER 9 REVIEW/PROBLEMS AND PUZZLES 288

CHAPTER 10 • WHAT EARTH PROVIDES 289

TOPIC 4: ROCKS AND SOIL: TWO RESOURCES 290
- **EXPLORE ACTIVITY** Investigate What Is in Soil 291
- **SKILL BUILDER** Measuring: Finding the Volume of a Water Sample 295
- **SCIENCE MAGAZINE** George Washington Carver: The Farmers' Friend 298

TOPIC 5: OTHER NATURAL RESOURCES 300
- **EXPLORE ACTIVITY** Investigate How Mining Affects Land 301
- **QUICK LAB** Energy Survey 304
- **SCIENCE MAGAZINE** Positively Plastic 306

TOPIC 6: CONSERVING EARTH'S RESOURCES 308
- **EXPLORE ACTIVITY** Investigate What Happens When Materials Get into Water 309
- **QUICK LAB** Cleaning Water 311
- **SCIENCE MAGAZINE** Pollution at Paint Creek 316

CHAPTER 10 REVIEW/PROBLEMS AND PUZZLES 317
UNIT 5 REVIEW/PROBLEMS AND PUZZLES 318–320

REFERENCE SECTION

HANDBOOK R1
    MEASUREMENTS R2–R3
    SAFETY R4–R5
    COLLECT DATA R6–R10
    MAKE MEASUREMENTS R11–R17
    MAKE OBSERVATIONS R18–R19
    REPRESENT DATA R20–R23
    USE TECHNOLOGY R24–R26
GLOSSARY R27
INDEX R39

UNIT 5
ROCKS AND RESOURCES

CHAPTER 9
THE CHANGING EARTH

Rocks are everywhere. They are on mountains and beaches. They are on the bottoms of rivers and at the bottom of oceans. Rocks have been around for a long time. Do rocks change? Has Earth's land always looked the way it does now?

In Chapter 9, you will read for cause and effect. Cause and effect helps you to understand both what happens and why it happens.

Topic 1
EARTH SCIENCE

WHY IT MATTERS

Rocks have many different uses.

SCIENCE WORDS

mineral a substance found in nature that is not a plant or an animal

landform a feature on the surface of Earth

plain a large area of land with few hills

valley an area of land lying between hills

plateau a flat area of land that rises above the land that surrounds it

Looking Under Your Feet

What was Earth like when dinosaurs roamed the land? Clues are right under your feet. The clues are rocks. Big ones like boulders. Little ones like grains of sand. Hard ones. Soft ones. Some that sparkle. Others that are dull or gray. Each kind of rock tells a different story.

EXPLORE

HYPOTHESIZE What makes up rocks? Are all rocks alike? Write a hypothesis in your *Science Journal*. How might you test your ideas?

258

EXPLORE ACTIVITY

Investigate How Rocks Are Alike and Different

Observe different rocks to infer how they are alike and different.

MATERIALS
- several different rocks
- hand lens
- *Science Journal*

PROCEDURES

1. **OBSERVE** Touch each rock. How does it feel? Record your observations in your *Science Journal*.

2. **OBSERVE** Look at each rock. Write or draw what you see. Write about any lines or patterns you observe.

3. **COMPARE** Now use your hand lens to look at each rock again. Write about what you see with the hand lens that you could not see before.

CONCLUDE AND APPLY

1. **COMPARE AND CONTRAST** Did any rocks feel the same? Did any rocks have similar lines or patterns?

2. **IDENTIFY** When you observed the rocks with a hand lens, did you observe any quality that all the rocks have in common?

3. **INFER** Are the rocks made of one material or more than one material? How can you tell?

GOING FURTHER: Apply

4. **CLASSIFY** Based on your observations, what are some ways the rocks could be grouped together to show their similarities or differences?

How Are Rocks Alike and Different?

The Explore Activity demonstrates that rocks are not all alike. In fact, there are many different kinds of rocks. Rocks are found in different sizes, shapes, and colors. Some rocks are smooth. Other rocks are rough.

Although rocks look and feel different from each other, all rocks are made of the same material. All rocks are made of **minerals** (min′ər əlz). A mineral is a substance found in nature that is not a plant or an animal. Minerals are the building blocks of rocks.

Some rocks, like granite (gran′it), are made up of many different minerals. Other rocks, like limestone (līm′stōn′), are made up of mostly one mineral. Minerals make a rock look and feel the way it does. Some minerals are hard. Other minerals are soft.

There are over 3,000 minerals and about 600 kinds of rocks.

mica
(mī′kə)

feldspar
(feld′ spär′)

quartz
(kworts)

Granite is made up of many different minerals.

Limestone is made up mostly of the mineral calcite.

QUICK LAB

Mineral Scratch Test

HYPOTHESIZE Which of these minerals do you think is the hardest: quartz, calcite, or mica? Write a hypothesis in your *Science Journal*.

MATERIALS
- large pieces of quartz, calcite, and mica
- penny
- iron nail
- *Science Journal*

PROCEDURES

1. **PREDICT** What do you think will happen if you scratch each mineral with your fingernail? The penny? The iron nail? Write your predictions in your *Science Journal*.

2. **OBSERVE** Test your predictions. Scratch each mineral with your fingernail, the penny, and the iron nail. Record your observations.

CONCLUDE AND APPLY

1. **INTERPRET DATA** List the minerals you tested in order of hardness, from softest to hardest. How did you determine the order?

2. **COMMUNICATE** Use your observations to create a chart that describes the three minerals you tested. Include drawings of the minerals.

Brain Power

Eyeglasses can be made with either plastic or glass lenses. Glass lenses have the mineral quartz in them. Plastic lenses are light, but can be scratched more easily. What might be the advantage of quartz-glass lenses?

How Are Rocks Formed?

All of Earth's rocks are formed in one of three ways.

Some rocks are formed when melted rock below the surface of Earth cools and hardens. Granite is formed in this way. Sometimes the melted rock flows out onto Earth's surface, where it cools and hardens more quickly. Rocks such as basalt (bə solt′) and obsidian (ob sid′ē ən) form in this way.

Other rocks are formed when bits of soil, mud, and rock in the bottoms of rivers, lakes, and oceans pile up over time. As the layers build up, the materials on the bottom of the pile begin to change. The layers at the bottom of the pile get cemented together, forming solid rock. Rocks such as conglomerate (kən glom′ər it), sandstone (sand′stōn′), and shale (shāl) are formed this way.

Limestone is another kind of layered rock. It is made up of shells, skeletons of tiny sea animals, and hardened parts of small sea plants.

Basalt

Limestone

Sometimes rocks get squeezed and heated below Earth's surface. When this happens, the rock's properties change. It becomes a new kind of rock. This is a third way new rocks are formed.

All three kinds of rocks can be changed over and over again. They form, break up, and reform. Old rocks become new rocks. The chart below gives some examples.

ROCKS: OLD AND NEW

Old	Squeezed and Heated	New
shale 	→	slate 
limestone 	→	marble 
slate 	→	schist 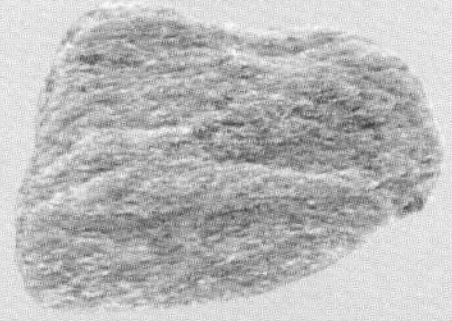
granite 	→	gneiss 

READING CHARTS

1. **WRITE** What are the names of the new rocks in the chart? Make a list.
2. **WRITE** Which new rock in the chart is also listed as an old rock? What new rock can it form?

What Is Earth's Surface Like Where You Live?

Earth is an enormous ball with an outer "crust" of rock. In some places the rock is exposed, like on the side of a mountain. In other places the rock might be covered by soil, sand, buildings, pavement, or water.

The surface of a ball is smooth, but the surface of Earth is not smooth at all! It has many different shapes. What shape does Earth's surface have where you live?

The different shapes on Earth's surface are **landforms** (land′formz′). A landform is a feature on the surface of Earth.

The land along the edge of an ocean or other body of water is a beach. Beaches are flat and narrow stretches of land. They are made up of sand, gravel, or pebbles. Beaches are one kind of landform.

Another landform is a **plain** (plān). A plain is a large area of land with few or no hills. Plains have thick layers of soil.

A third kind of landform is a **valley** (val′ē). A valley is an area of land lying between hills or mountains. Rivers or streams often flow through the lowest parts of valleys.

Hills are rounded and raised landforms. They are higher than plains and valleys, but lower than mountains. Mountains are the highest landforms. Most mountains have large areas of exposed rock. In some places the rock may be covered by a thin layer of soil.

Beach

Valley

Plains

WHY IT MATTERS

Whether you walk up a mountain or down into a valley, you are walking over rock. Rocks are always under your feet, but sometimes they are over your head, too! People use rocks to build many of the bridges, buildings, statues, and walls that you see every day.

This landform is a plateau (plātō′). A plateau is a flat area of land that rises above the land that surrounds it.

REVIEW

1. Name and describe three landforms. What lies under each of these landforms?

2. What are rocks made of?

3. Name one mineral and describe its characteristics.

4. **COMMUNICATE** Describe three ways rocks are formed.

5. **CRITICAL THINKING** *Evaluate* How are mountains, hills, and plateaus alike? How are they different?

WHY IT MATTERS THINK ABOUT IT What kinds of rocks have you seen in your community? Describe their colors, sizes, and shapes.

WHY IT MATTERS WRITE ABOUT IT How are rocks used to make structures in your community? Describe some structures made from rock that you've seen. Where did you see them?

NATIONAL GEOGRAPHIC World of SCIENCE

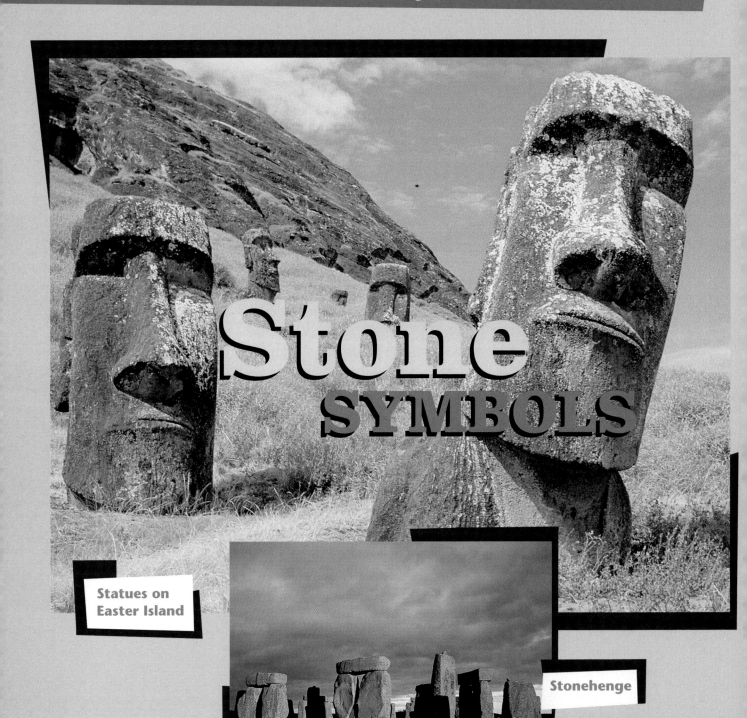

Stone SYMBOLS

Statues on Easter Island

Stonehenge

Geography Link

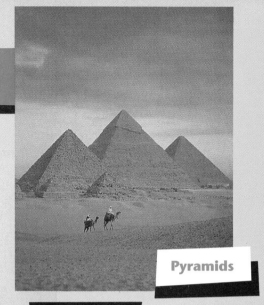
Pyramids

Long ago, people began building with rock. Many old structures stand today as monuments. They help us learn about ancient cultures.

Stonehenge is one of the world's oldest monuments. It was built in England about 5,000 years ago. Huge sandstone blocks and smaller rocks are arranged in circles. The largest rocks weigh 50 tons! Ancient people may have used Stonehenge to mark the movements of the Sun.

Great Wall

About 4,500 years ago, the people of Egypt built large stone pyramids as tombs for their kings. Millions of limestone blocks were used for the pyramids. Pyramids point to the sky. Egyptians may have thought the pyramids helped their kings reach heaven!

The Great Wall of China was built more than 2,000 years ago to keep out invaders. It's made mostly of granite and brick. At about 6,400 kilometers (4,000 miles) in length, the Great Wall is the longest structure ever built!

Giant statues stand on Easter Island, west of Chile. They were carved from volcanic rock more than 1,000 years ago. Hundreds of statues have been found, some as tall as 22 meters (66 feet). They may have been monuments to people who had died.

Discussion Starter

1. Name a stone monument in our country.

2. Which American monuments do you think will best tell future people about our culture?

*inter*NET CONNECTION To learn more about stone monuments, visit www.mhschool.com/science and enter the keyword **STATUES.**

Topic 2
EARTH SCIENCE

WHY IT MATTERS

You can see changes to Earth's surface all around you.

SCIENCE WORDS

weathering the process that causes rocks to crumble, crack, and break

erosion occurs when weathered materials are carried away

glacier a large mass of ice in motion

Slow Changes

How can a chunk of rock be turned into a soaring tower? How can a giant boulder be changed into a beautiful bridge? Is it magic? No! Something else is at work. How do you think these features were formed?

EXPLORE

HYPOTHESIZE Chalk is a kind of rock made up of the mineral calcite. How do you think the chalk got its shape? What are some ways you could change its shape? Write a hypothesis in your *Science Journal*. How might you test your ideas?

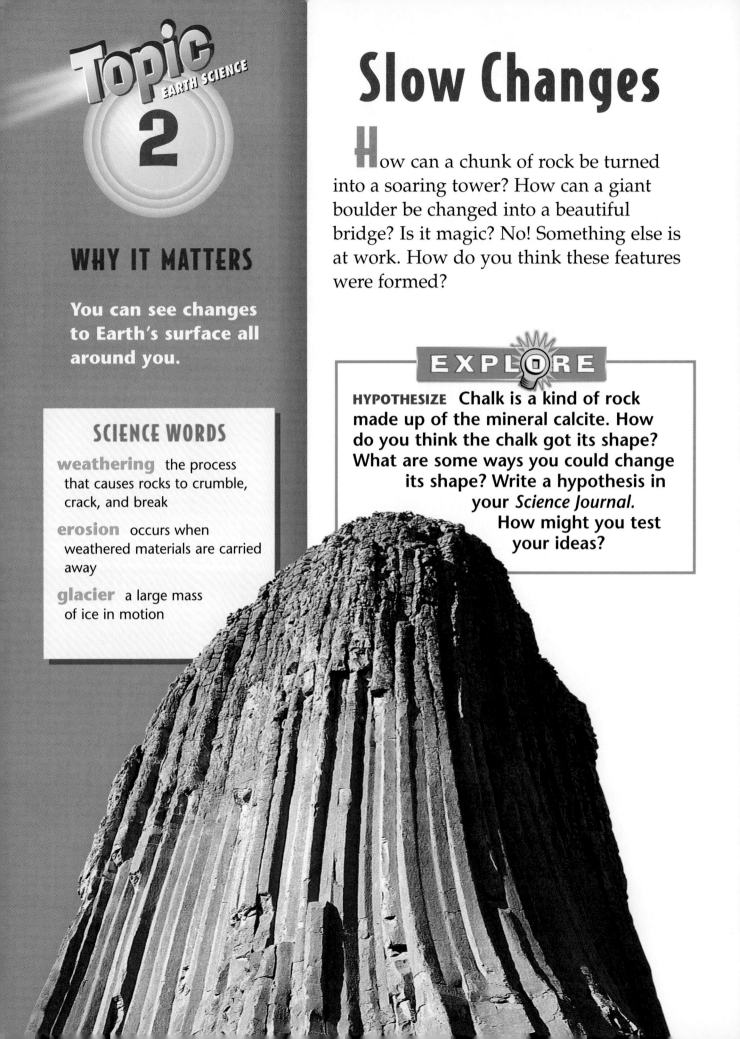

EXPLORE ACTIVITY

Investigate How Rocks Change

Investigate how rocks can change. Test what happens to chalk when you write on sandpaper.

MATERIALS
- 3 pieces of chalk (different colors)
- piece of sandpaper
- *Science Journal*

PROCEDURES

1. **OBSERVE** Look at your piece of chalk and your sandpaper. Write down what they look and feel like in your *Science Journal*. List as many properties of each one as you can.

2. **PREDICT** What will happen when you draw with the chalk on the sandpaper? Write down your prediction.

3. Draw with your chalk on the sandpaper. You might draw a picture or write some words.

4. **COMPARE** Look at your chalk. How does the way it looks and feels now compare to before? Look at your sandpaper. How does the way it looks and feels now compare to before?

CONCLUDE AND APPLY

1. **DESCRIBE** What happened to the piece of chalk? What happened to the sandpaper?

2. **EXPLAIN** Which material changed the most? Why do you think this happened?

GOING FURTHER: Apply

3. **INFER** Strong winds sometimes carry sand as they blow over rocks. How might the wind and sand change the rocks over time?

How Do Rocks Change?

All rocks are eventually broken up. One process that helps to break up rocks is **weathering** (weth′ər ing). Weathering is the process that causes rocks on Earth's surface to crumble, crack, and break. Things such as chemicals, water, temperature changes, plant roots, ice, and wind can cause the weathering of rocks. Weathering is usually a very slow process.

The Explore Activity demonstrates how a piece of chalk changes shape when it is rubbed on a rough surface. Wind carrying sand and pebbles can blast the surface of rocks, wearing the rocks down over time.

Water causes the weathering of rocks on beaches and riverbeds. Over time, these rocks also change shape and break down. When water gets into spaces in and around rocks and then freezes, rocks are broken open. Plant roots can do this, too.

Chemicals (kem′i kəlz) can also change rocks. Chemicals are all around you. They are in the air and water. When rocks are exposed to chemicals, the minerals in the rocks soften. Then the rocks change shape.

Ice can break rocks apart.

HOW A CAVE FORMS

Water runs down through cracks in limestone.

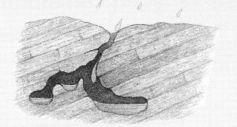

Chemicals in the water soften the minerals in the rock. A small opening forms.

The space gets larger. A cave has formed.

READING DIAGRAMS

1. **DISCUSS** What softens the minerals in the rock?
2. **WRITE** What is the large space in the rock called?

Changing Chalk

HYPOTHESIZE Vinegar is a kind of chemical. What will happen when chalk is put in vinegar? Write a hypothesis in your *Science Journal*.

MATERIALS
- apron
- safety goggles
- piece of chalk
- small jar
- vinegar
- *Science Journal*

PROCEDURES

 SAFETY: Wear Goggles.

1. Put on your apron and safety goggles. Place the chalk in the jar and pour in just enough vinegar to cover the chalk.

2. **OBSERVE** After a few seconds, describe what you see. Record your observations in your *Science Journal*.

CONCLUDE AND APPLY

EXPLAIN Why did the chalk change? How do you know?

What Happens to Weathered Materials?

Weathered materials don't stay put. They are moved around by **erosion** (i rō′zhən). Erosion occurs when weathered materials are carried away. Erosion, like weathering, is usually a slow process. Ice, water, gravity, and wind all help to move weathered materials around.

Erosion happens all around us. Rivers and streams carry weathered materials over great distances. Gravity can pull weathered materials down hills and mountains. Winds pick up and move large amounts of soil and sand.

Glaciers (glā′shərz) cause erosion, too. A glacier is a large mass of ice in motion. If a glacier is up high, it will move to lower land. If it is on flat land, it will begin to spread. As glaciers move over the land they pick up weathered materials and rocks. These materials become stuck in the glacier, scraping the ground.

Glaciers form when more snow falls than can melt away. As the snow piles up, it eventually begins to move.

This large pile of rocks formed when gravity pulled rocks down a hillside.

SKILL BUILDER

Skill: Forming a Hypothesis

WHICH MATERIALS SETTLE FIRST?

What happens to the material that is carried away by erosion? In this activity, you will put pebbles, sand, soil, and water in a jar. After you shake the jar, you will let it sit for several hours. How will the materials in the jar look after several hours? Write down what you think the outcome will be. This is your hypothesis. A hypothesis is an answer you propose to a question that can be tested.

MATERIALS
- a large plastic jar with lid
- measuring cup
- pebbles
- sand
- soil
- water
- *Science Journal*

PROCEDURES

1. Put one cup of each of the materials (pebbles, sand, and soil) in the jar and fill the jar with water. Put the lid on the jar and shake it.

2. **OBSERVE** Write a description of what your jar looks like in your *Science Journal*. Draw a picture of the jar and the materials.

3. **OBSERVE** Let the jar sit for several hours, then observe the contents again. How does the jar look now?

CONCLUDE AND APPLY

1. **COMPARE** How did the jar look when you first observed it? How did it look several hours later? Was your hypothesis correct?

2. **DESCRIBE** Which material in the jar settled first? Last? How do you know?

3. **INFER** A fast-moving stream may carry pebbles, sand, and soil. As the water slows down, which material will settle out first? Which material will be carried the greatest distance?

How Do Things in Your Neighborhood Change?

No matter where you live, you can observe some of the changes brought about by weathering and erosion. As plants grow, their roots may break apart sidewalks, or roads. Changes in temperature, and the action of chemicals, water, and ice also cause sidewalks and pavement to bend, crack, and crumble. What other examples of changes caused by weathering and erosion have you observed in your community?

What caused the words on this grave marker to disappear?

Brain Power

Tree roots pushing up from below have caused this sidewalk to break. The tree has changed its environment. In what other ways do living things change their environment to meet their needs?

WHY IT MATTERS

Weathering and erosion are always at work. You can see the changes they cause all around you. People make changes to Earth's surface, too. Farmers move rocks and soil to prepare fields for planting. Hikers clear trails in the woods. Families plant gardens and children pick up rocks to collect. Homes, towns, and cities are built.

Rocks are not the only materials affected by weathering. This weathered house was once a solid wooden structure covered by paint.

REVIEW

1. How do rocks change?
2. How is weathering different from erosion?
3. Name three things that cause weathering. Name three things that cause erosion.
4. **HYPOTHESIZE** Devil's Tower, shown on page 268, is the "neck" of an ancient volcano. Why are the rocks that once surrounded this feature no longer there? State a hypothesis.
5. **CRITICAL THINKING** *Evaluate* Water can cause both weathering and erosion. What properties does water have? How do these properties help water to cause so much change?

WHY IT MATTERS THINK ABOUT IT
Think about your schoolyard. What evidence of changes have you observed there? What changes were made by people?

WHY IT MATTERS WRITE ABOUT IT
How might your schoolyard look ten years from now? Twenty years from now?

SCIENCE MAGAZINE

ERASING

Soil is very important, especially on farms. Crops need rich soil to help them grow. Wind and rain can wash away, or erode, farm soil. Farmers try to slow erosion and protect their farmland.

Contour (kon'tur) farming protects the soil, too. Crops are planted in rows around hills. Each row soaks up rainwater as it runs down the hill.

Terrace (ter'is) farming protects the soil. "Shelves" are cut into the sides of a hill. The flat shelves hold rainwater better than a steep hillside!

Cover crops can help save soil. These crops aren't for harvesting. They're planted in empty fields. Their roots hold the soil in place during storms.

Earth Science Link

EROSION

Rows of trees can be planted between crop fields. The trees help break the force of the wind. Gentle winds don't carry as much soil away as strong winds.

During the 1930s the Great Plains were very dry. Dust storms carried away loose soil. Now farmers plant trees and terrace hillsides to keep the soil in place.

In Texas, people fight erosion in two ways. They reseed the prairies and terrace the farmland in hilly fields.

DISCUSSION STARTER

1. What ways of controlling erosion have you seen in your community?
2. Compare and contrast contour farming and terrace farming.

To learn more about erosion, visit *www.mhschool.com/science* and enter the keyword WEAROFF.

*inter*NET CONNECTION

Topic 3
EARTH SCIENCE

WHY IT MATTERS

It's important to be prepared for fast changes to Earth's surface.

SCIENCE WORDS

hurricane a violent storm with strong winds and heavy rains

earthquake a sudden movement in the rocks that make up Earth's crust

volcano an opening in the surface of Earth

Fast Changes

How is a storm like a broom? Can you solve this riddle? Sometimes land can change in just a few hours. One day it's here. Another day it's somewhere else. You may think it's gone altogether. However, it has just been moved around.

Think about what you know about weathering and erosion. What kinds of changes might this storm cause?

EXPLORE

HYPOTHESIZE A hurricane is a powerful storm. A drizzle isn't a storm at all. Each dumps water on land. Which kind of rainfall causes the most erosion? Write a hypothesis in your *Science Journal*. How might you test your ideas?

EXPLORE ACTIVITY

Design Your Own Experiment

HOW CAN LAND CHANGE QUICKLY?

PROCEDURES

1. Mix some rocks and soil together. You can smooth the rocks and soil or shape them into a hill, but don't pack them down. Draw how your "land" looks in your *Science Journal*.

2. **MODEL** How can you use cups A and B to model a gentle rain and a heavy rain? How can you use the pencils to help you?

3. **EXPERIMENT** How is your land affected by a gentle rain? How is your land affected by a heavy rain?

MATERIALS
- 2 paper cups, labeled A and B
- 2 trays, labeled A and B
- 2 different-sized pencils
- measuring cup
- soil and rocks
- water
- *Science Journal*

CONCLUDE AND APPLY

1. **COMPARE** What happened to the two trays? In which tray did the most change take place?

2. **DRAW CONCLUSIONS** Which kind of rainfall causes the most erosion? How do you know?

GOING FURTHER: Problem Solving

3. **EXPERIMENT** How could you test the effects of wind on land?

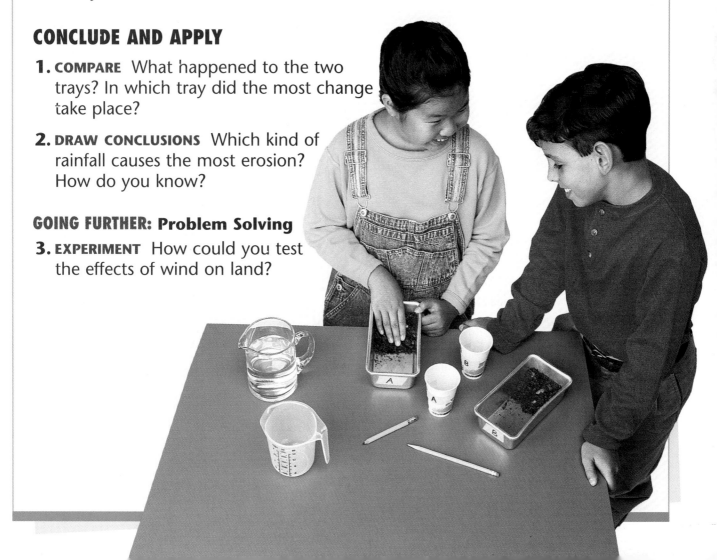

How Can Land Change Quickly?

Weathering and erosion usually take place slowly over long periods of time. The Explore Activity demonstrates that a heavy rainstorm can cause erosion quickly. Heavy rainstorms speed up the weathering and erosion of land.

When weathering and erosion happen quickly, there are sudden changes to Earth's surface. There are several events that can speed up the weathering and erosion of Earth's land. One of these events is a **hurricane** (hur′i kān′).

A hurricane is a violent storm with strong winds and heavy rains. Hurricanes are the largest and most powerful of all storms. They begin over the oceans. Hurricane winds move in a circular pattern at speeds of 75 miles an hour or more!

Brain Power

Hurricanes are just one kind of violent event that cause sudden changes to Earth's surface. Can you think of any others?

Most hurricanes die out far from land. Hurricanes that move toward land act like giant bulldozers. Strong winds and giant waves damage or destroy almost everything in their paths. Coasts can be changed—often in minutes. Flooding occurs. Houses, roads, bridges, and cars may be swept away. Trees are uprooted. Lives are often lost.

Hurricane damage

Weather Adds Up

HYPOTHESIZE Some cities have a lot of rain each year. Other cities only have a little. In which kind of city will a stone monument change more quickly? Write a hypothesis in your *Science Journal*.

MATERIALS
- Science Journal

PROCEDURES

1. **INTERPRET DATA** Look at the table. What is the difference in winter temperatures between the two cities? Write the number in your *Science Journal*.

2. **INTERPRET DATA** Which city has more rainfall?

CONCLUDE AND APPLY

DRAW CONCLUSIONS A stone monument was moved to New York City from Egypt. The monument has changed very quickly in New York. Why do you think this has happened?

COMPARING WEATHER IN NEW YORK AND CAIRO

	New York City, USA	Cairo, Egypt
Average Temperature in Winter	32°F	56°F
Average Rainfall Each Year	107 cm	0 to 10 cm

What Other Events Cause Sudden Changes to Earth's Surface?

Another event that causes sudden changes to Earth's surface is an earthquake (urth'kwāk'). An earthquake is a sudden movement in the rocks that make up Earth's crust. Earthquakes are caused by forces deep within Earth. The forces cause the rocks to break. The breaking rock makes the ground shake.

Some earthquakes are so weak they can hardly be felt. Others are very strong. Strong earthquakes cause landslides, great destruction, and loss of lives. People may be left homeless when houses are destroyed. Cities may lose electricity and water as power and water lines break. Food supplies may run low if transportation is interrupted. Fires may also break out.

Landslides often happen when earthquakes do. They also occur when heavy rains or melting snow loosen rocks, soil, sand, and gravel. Then mud, sand, and boulders tumble down from mountains and hills. Buildings may be damaged, swept away, or buried.

earthquake damage

landslide damage

HOW A VOLCANO FORMS

Lava and other materials flow out onto the surface of Earth during a volcanic eruption.

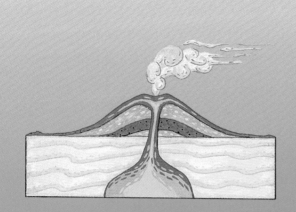

The materials pile up around the opening, forming a volcanic mountain, or volcano.

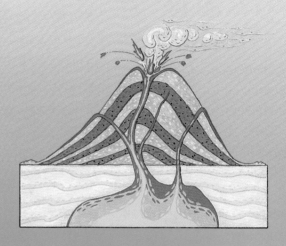

READING DIAGRAMS

1. **WRITE** What flows out of a volcano?
2. **REPRESENT** How would the volcano look after another eruption?

Sudden changes to Earth's surface are also caused by **volcanoes** (vol kā′nōz). A volcano is an opening in the surface of Earth. Melted rock, gases, pieces of rock, and dust are forced out of this opening. The word *volcano* is also the name of the landform that is built up around this opening. Melted rock that flows out onto the ground is called lava (lä′və).

Volcanoes may not erupt for hundreds of years and then erupt again. A large disaster happened in Italy about 2,000 years ago when Mt. Vesuvius (və sü′vē əs) erupted. Many people died from breathing deadly gases. Three entire cities were buried under lava, ash, and volcanic rocks. Hundreds of years later, the remains of these cities were discovered.

NATIONAL GEOGRAPHIC
FUNtastic Facts

When Krakatoa, a volcanic island in the South Pacific, exploded in 1883, the blast was heard in Perth, Australia, nearly 2,000 miles away! What does that tell you about the explosion?

HISTORY LINK
What Is the Dust Bowl?

Terrible dust storms spelled disaster for many people living in the Great Plains of the United States in the 1930s. This area of the country became known as the Dust Bowl.

Farmers in the Great Plains dug up millions of acres (ā′kərz) of land in order to grow crops. For many years there was plenty of rain. There were also very good harvests.

Then, in 1931, a long, dry period began. Crops did not grow and the soil dried out. Without plants to hold the soil in place, the dry soil began to blow away with the wind. Thick, black clouds of dust were carried hundreds of miles across the plains. When the dust settled, it buried fences, houses, and farm buildings.

As the dust storms continued over the next several years, more and more farmers had to move away. Finally, eight years later, the rains returned. A hard lesson had been learned. Farmers had learned to plant crops that would protect the soil from erosion.

WHY IT MATTERS

Sudden events like hurricanes, earthquakes, landslides, and volcanic eruptions can cause a lot of damage. People cannot prevent these events from happening, but they can prepare for them. Weather forecasters can predict when hurricanes will move toward land. Scientists can predict when earthquakes and volcanic eruptions are likely to occur.

REVIEW

1. How can land change quickly?
2. Name two things that work to change the shape of the land when hurricanes hit the coast.
3. What is a volcano?
4. **COMMUNICATE** What caused the dust storms of the 1930s? Draw a diagram to explain. Label your diagram.
5. **CRITICAL THINKING** *Apply* Describe some of the ways people are affected when strong earthquakes occur where they live.

WHY IT MATTERS THINK ABOUT IT
Describe the emergency drills you have at your school. How should you act in an emergency?

WHY IT MATTERS WRITE ABOUT IT
Why is it important to know what to do in the case of an emergency?

READING SKILL How does a volcano form? Write a paragraph that explains the cause and the effect.

SCIENCE MAGAZINE

PREDICTING HURRICANES

When Christopher Columbus arrived in the New World, he was met by Native Americans. They told him about storms they called hurricanes. They said the storms brought great winds, heavy rains, and giant waves. They never knew when a hurricane would form or where it would strike land!

Today four modern tools help weather forecasters predict when and where a hurricane will strike. This information gives people time to board up their windows and find a safe place to stay.

Brave pilots fly planes into the center, or eye, of a hurricane. Instruments on the planes measure the hurricane's size and strength, and what direction it's moving.

Science, Technology, and Society

DISCUSSION STARTER

1. What tools are used to predict a hurricane?
2. How would you prepare for a hurricane?

A Weather satellites are always orbiting Earth. They take pictures of weather patterns around the world. Satellite pictures can show hurricanes forming and crossing the Atlantic Ocean.

B Radar sends out radio signals that bounce back off raindrops. The radar screen can show a hurricane's heavy rains when they are still very far away.

C Denise Stephenson-Hawk is a scientist who studies the atmosphere. She uses models to learn how conditions in the atmosphere, oceans, and land may be used to predict the weather.

To learn more about predicting weather, visit www.mhschool.com/science and enter the keyword STORMY.

interNET CONNECTION

CHAPTER 9 REVIEW

SCIENCE WORDS

earthquake p.282
erosion p.272
glacier p.272
hurricane p.280
landform p.264
mineral p.260
plain p.264
plateau p.265
valley p.264
volcano p.283
weathering p.270

USING SCIENCE WORDS

Number a paper from 1 to 10. Fill in 1 to 5 with a word from the list above.

1. An opening in Earth's crust through which melted rock and other materials are forced out is a __?__.

2. All rocks are made up of __?__.

3. The process that causes rocks to crack, crumble and break is __?__.

4. A flat area of land that rises above the land that surrounds it is called a __?__.

5. A sudden movement in the rocks that make up Earth's crust is called an __?__.

6–10. Pick five words from the list above that were not used in 1 to 5 and use each in a sentence.

UNDERSTANDING SCIENCE IDEAS

11. Beaches often have laws against walking on sand dunes. Why do you think this is so?

12. How does a glacier cause erosion?

USING IDEAS AND SKILLS

13. **READING SKILL: CAUSE AND EFFECT** On a walk you see an old, crumbling stone wall. Why is the wall falling apart?

14. **HYPOTHESIZE** How do you think the three landforms shown below are related? Write a hypothesis.

Mesa

Plateau

Butte

15. **THINKING LIKE A SCIENTIST** You find a rock near your school. How could you test how hard it is?

PROBLEMS and PUZZLES

Lava Flow Can you make a model of a volcano? Wear goggles. Fill half a bottle with warm water. Add a few drops of liquid detergent and vinegar. Add a teaspoon or two of baking soda and stand back!

CHAPTER 10
WHAT EARTH PROVIDES

You walk on it. You run on it. Your school is built on it. Earth is your home! Have you ever thought about all the things that Earth provides? What kinds of things do you take from Earth? What do you put back in their place?

In Chapter 10 you will read for the order, or sequence, of events. Knowing the sequence of events helps you to understand and remember them.

Topic 4
EARTH SCIENCE

WHY IT MATTERS

Rocks and soil are necessary to your community.

SCIENCE WORDS

natural resource
a material on Earth that is necessary or useful to people

renewable resource
a resource that can be replaced or used over and over again

Rocks and Soil: Two Resources

Would these plants exist if there were no soil? Tall trees reach into the sky. Little garden plants brush your ankles. Tall or short, these plants have something in common. They have roots. Roots reach into the soil to soak up water and minerals. Have you ever looked closely at soil?

EXPLORE

HYPOTHESIZE You know that soil is important for growing plants. What is in soil? Write a hypothesis in your *Science Journal*. How might you test your ideas?

EXPLORE ACTIVITY

Investigate What Is in Soil

Examine a soil sample to find out what is in soil.

MATERIALS
- small amount of soil
- piece of white paper
- hand lens
- *Science Journal*

PROCEDURES

1. Spread out your soil sample on the piece of white paper.

2. **OBSERVE** Look at the soil closely. What do you see? Are there different colors? Different size particles? Write your observations in your *Science Journal*.

3. **OBSERVE** Smell your soil. How does it smell? Feel your soil. How does it feel? Are there different textures? Write down a description of your soil based on how it looks, feels, and smells. Be sure to wash your hands after touching the soil.

4. **COMPARE** Using your hand lens, examine your soil again. What do you see that you didn't see before? Record your new observations.

CONCLUDE AND APPLY

1. **COMMUNICATE** What did you see in your soil?

2. **IDENTIFY** Was there anything in the soil that surprised you? What was it?

GOING FURTHER: Apply

3. **HYPOTHESIZE** Think about what you saw in your soil. How do you think those things got into the soil?

Why Are Rocks and Soil Important?

Rocks and soil are **natural resources** (nach′ər əl rē′sors′əz). A natural resource is a material on Earth that is necessary or useful to people.

How are rocks useful to people? You know that rocks are important building materials. Crushed limestone is used in making cement. Some roads are covered with sand or gravel. Sand is also an important ingredient in glass. Clay and sand are used in making bricks. Weathered rocks are part of the soil. If there were no rocks, there would be no soil.

Soil is the part of the ground that plants grow in. Without soil, there would be no plants on the land. People and animals need plants for food.

Deer like to eat small branches and leaves from trees or bushes.

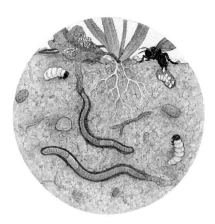

What materials make up soil?

What Is in Soil?

The Explore Activity shows that soil is made up of several different materials. Soil is a mixture of tiny rock particles, minerals, and decayed plant and animal material. Growing plant roots, worms, and insects make spaces for air and water.

There are usually two layers of soil. The top layer, called topsoil, is made up of very small particles that are dark in color. Topsoil has a lot of minerals and decayed plant and animal material. These things are necessary for plant growth. Topsoil holds water well, which is also necessary for plant growth.

A layer of soil called subsoil lies below the topsoil. Subsoil is made up of bigger particles and is lighter in color than topsoil. It does not have decayed plant and animal material, but it does hold water and has some of the minerals needed by plants. Below the subsoil is solid rock.

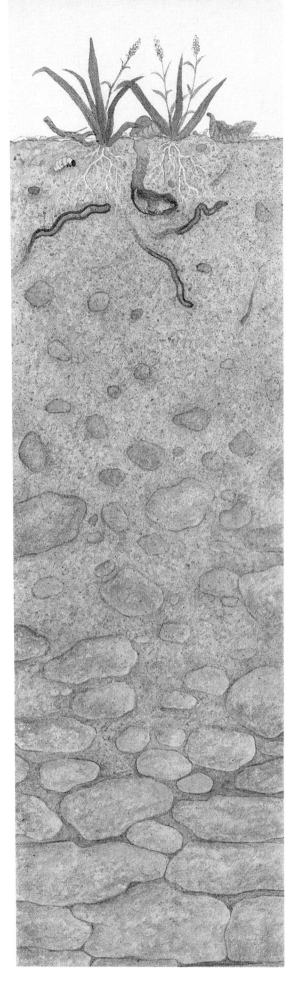

topsoil

subsoil

solid rock

Are There Different Kinds of Soils?

There are many different kinds of soils. Different soils have different types of rock and minerals in them. Some soils have more water in them than others. Some soils might have more plant and animal material in them, too.

Different kinds of soils are found in different parts of the world. There are several kinds of soils found in the United States. In some areas, the soil has a lot of clay. Other soils are very sandy. Loam (lōm) is a kind of soil that has a good mixture of clay and sand.

In some places, soil layers are very thick. Lots of plants grow in places with a thick soil layer. In dry and windy places soil layers are much thinner. Layers of soil on mountains are thin because gravity pulls the soil downhill.

The type of soil in a particular place affects what kinds of plants can grow there.

This soil is sandy.

Plants grow well in loam.

This soil has a lot of clay in it.

SKILL BUILDER

Skill: Measuring

FINDING THE VOLUME OF A WATER SAMPLE

The amount of water that soil can hold is an important property of soil. In this activity, you will measure the amount of water held by two soil samples. Sometimes you measure something to find out an object's size, weight, or temperature. In this activity, you will measure the volume of a sample of water. You can use a calculator to help.

MATERIALS
- 4 paper cups, 2 with holes
- measuring cup
- water
- sandy soil
- potting soil
- graduated cylinder
- watch or clock
- calculator (optional)
- *Science Journal*

PROCEDURES

1. **MEASURE** Using the measuring cup, measure 1 cup of potting soil into a paper cup with holes. Pack the soil down. Label the cup. Measure 100 mL of water in the graduated cylinder.

2. **EXPERIMENT** Hold the paper cup with potting soil over a paper cup without holes. Have your partner pour the water slowly into the cup of soil. Let the water run through the cup for two minutes. If there is still water dripping out, place the cup of soil inside an empty cup.

3. **MEASURE** Pour the water that ran out of the cup into the graduated cylinder. Read and record the volume of water in the graduated cylinder. Write the number in your *Science Journal*.

4. **EXPERIMENT** Repeat steps 1–3 with the sandy soil.

CONCLUDE AND APPLY

1. **INTERPRET** Which soil held more water? How do you know?

2. **APPLY** Which type of soil would be best for use in a garden? Which sample would be best for use on a soccer field?

Why Is It Important to Conserve Natural Resources?

Some natural resources are **renewable resources** (ri nü′ə bəl rē′sors′əz). A renewable resource is a resource that can be replaced or be used over and over again. Soil is an example of a renewable resource.

However, the erosion of soil happens quickly when soil is left unprotected. Soil is unprotected when forests are cut down and when farmland is left bare. You read about how this happened in the Dust Bowl on page 284.

Although soil is renewable, it takes many, many years for it to form. For this reason, it is important to conserve soil. To conserve something means to protect it and use it wisely. Good farming practices, like contour farming and planting cover crops, help conserve the soil.

Soybeans are planted as a cover crop on this field.

Brain Power

Some Native American peoples only took a plant for use if they could find several others just like it. Why do you think they did this?

WHY IT MATTERS

The next time you take a walk, take a good look around you. How does the land look where you live? The way the land looks depends on what rocks and soil are at Earth's surface. The area where you live feels like home to you. Rocks and soil are part of what makes your home special.

REVIEW

1. What is a natural resource?
2. Why is soil a natural resource?
3. In what ways do soils differ?
4. **MEASURE** Every time you water your classroom plant, water drips onto the floor. How can you find out how much water the plant is losing?
5. **CRITICAL THINKING** *Apply* You know that soil is important to plants. Do you think plants are important to soil? Explain.

WHY IT MATTERS THINK ABOUT IT
Describe the area of the country where you live. Why do you like it?

WHY IT MATTERS WRITE ABOUT IT
How is the place where you live different from other places? What makes it special to the people who live there?

SCIENCE MAGAZINE

George Washington Carver: The FARMERS' FRIEND

History of Science

Next time you eat a peanut, think of George Washington Carver: He's the man who made the peanut famous!

Carver was born in slavery in Missouri in 1861. Even at a young age, he was very interested in plants.

At age 12, Carver went away to get an education. By 1896 he had earned two college degrees. Then he got a job at the Tuskegee Institute in Alabama. His goal was to help poor Southern farmers.

Most Southern farmers grew only cotton. The plant took all the nitrogen (nī′trə jən) out of the soil. Carver suggested that they plant cotton one year and a different crop the next year. He wanted the farmers to use peanuts, soybeans, and sweet potatoes. Why? As these plants grow, they put nitrogen back into the soil!

Carver developed more than 300 products from peanuts alone! These included milk, coffee, flour, and soap. He also developed more than 100 products from sweet potatoes, including flour, rubber, and glue!

Soon many companies bought the new crops and made the new products. More and more Southern farmers rotated their crops with cotton. In time, their cotton grew better, too.

George Washington Carver received many honors. He was most proud of helping the farmers. They began to earn more money. Best of all, their farmland could support generations of farmers in the future.

DISCUSSION STARTER

1. Why did Carver work to develop products made from peanuts and sweet potatoes?

2. How did Carver prove that soil is a renewable resource?

To learn more about Carver's work, visit **www.mhschool.com/science** and enter the keyword CARVER.

*inter*NET CONNECTION

Topic 5
EARTH SCIENCE

WHY IT MATTERS

Earth provides many different resources.

SCIENCE WORDS

nonrenewable resource a resource that cannot be reused or replaced in a useful amount of time

Other Natural Resources

Are diamonds a natural resource? Diamonds are hard to find. They are hidden below Earth's surface. They must be mined, or dug up, before they can be used. It takes a lot of work to mine diamonds! It takes even more work to prepare a diamond so that it can be used.

Do you think diamonds are a renewable resource?

HYPOTHESIZE Some natural resources, such as diamonds and metals, are hard to find. What effect do you think mining these resources might have on Earth's land? Write a hypothesis in your *Science Journal*. How might you use a model to test your ideas?

EXPLORE ACTIVITY

Investigate How Mining Affects Land

The cookie in this activity represents an area of land. The chocolate chips in the cookie are the resource. Experiment to find out what you must do to get the resource.

MATERIALS
- chocolate chip cookie
- 4 toothpicks
- paper towel
- *Science Journal*

PROCEDURES

1. **OBSERVE** Place your cookie on the paper towel. Draw how your cookie looks in your *Science Journal*. Label your drawing.

2. **MODEL** Using your toothpicks, try to remove the resource (chocolate chips) from the land (cookie) without damaging the land. Before you begin, discuss with your partner how you will mine the resource.

3. **EXPERIMENT** Mine all the resource from your land. Draw how your cookie looks now.

CONCLUDE AND APPLY

1. **COMPARE** How does your cookie look after you removed all the chocolate chips?

2. **DRAW CONCLUSIONS** What happens when you can't find any more chips? How can you get more?

3. **INFER** What might be some problems people face in trying to mine resources from Earth?

GOING FURTHER: Problem Solving

4. How can damage to mining areas be repaired?

How Does Mining Affect Land?

The Explore Activity demonstrates that some resources take a lot of work to find. When one area of land has been mined of all of a resource, you must go to another place to find more. Mining often damages land.

Diamond

Diamonds take billions of years to form. Not only do they take a long time to form, but a lot of work is needed to find them. Diamonds are an example of a **nonrenewable resource** (non′ri nü′ə bəl rē′sors′). A nonrenewable resource is a resource that cannot be reused or replaced in a useful amount of time.

Many of the fuels we use for energy are also nonrenewable resources. Coal, oil, and natural gas are fuels. This means they can be used for energy.

About 200 years ago, most people in the United States burned wood for fuel. One hundred years ago, coal provided almost all the energy used in the United States. Today, oil and natural gas are the fuels used for most of our energy needs.

Coal, oil, and natural gas are nonrenewable resources. Coal is a hard, dark brown or black substance that is found in layers in the ground. It formed from the remains of plants that lived long ago.

Wood stove

HOW COAL FORMS

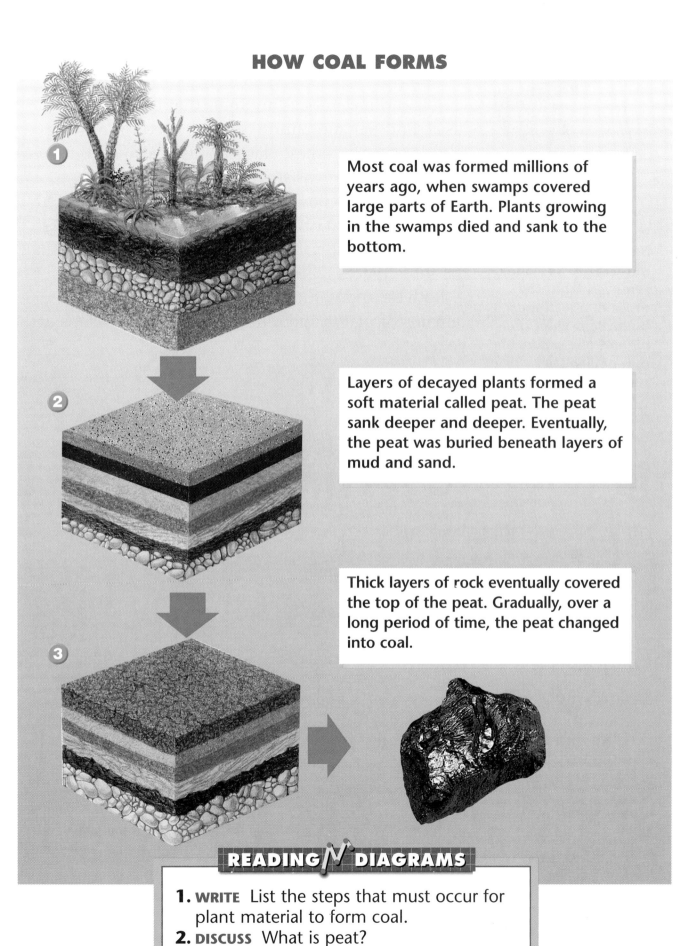

1. Most coal was formed millions of years ago, when swamps covered large parts of Earth. Plants growing in the swamps died and sank to the bottom.

2. Layers of decayed plants formed a soft material called peat. The peat sank deeper and deeper. Eventually, the peat was buried beneath layers of mud and sand.

3. Thick layers of rock eventually covered the top of the peat. Gradually, over a long period of time, the peat changed into coal.

READING DIAGRAMS

1. **WRITE** List the steps that must occur for plant material to form coal.
2. **DISCUSS** What is peat?

How Are Natural Gas and Oil Used?

Before oil and natural gas can be removed from the ground, a well must be drilled. Then the oil and gas can be pumped to the surface.

Oil and natural gas are fuels that have many uses. Both oil and natural gas are used for heating buildings and homes. Natural gas is also used for cooking food on gas stoves. Oil is made into gasoline, which powers cars and other vehicles. Oil is a thick, brown or black substance found in rocks below Earth's surface. Natural gas is often found in the same places that oil is found.

Both oil and natural gas formed from the remains of plants and animals that lived millions of years ago.

Energy Survey

HYPOTHESIZE Do you think more people in your class have a gas stove or an electric stove? Write a hypothesis in your *Science Journal*.

MATERIALS
- Science Journal

PROCEDURES

COLLECT DATA Conduct a survey of your classmates. Who has an electric stove? Who has a gas stove? Record the information in your *Science Journal*.

CONCLUDE AND APPLY

1. **IDENTIFY** How many people have electric stoves? How many people have gas stoves?

2. **COMMUNICATE** Make a bar graph that shows the results of your survey.

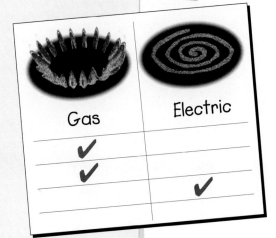

What Kind of Resources Are Water and Air?

All living things need water. Water can be used over and over again, but it is never used up. Water is a renewable resource.

Almost all water is found in Earth's oceans. Ocean water is salt water. Fresh water is found in glaciers, rivers and lakes.

Air is a mixture of gases. Air, like water, is necessary for living things. It is also a renewable resource.

WHY IT MATTERS

Like all living things, you need water and air to survive. People use coal, oil, and natural gas as fuels. You can survive without fuel, but your life would be very different and much less comfortable.

REVIEW

1. What is a nonrenewable resource?
2. How do we use coal, oil, and natural gas?
3. What kinds of resources are water and air? How do you know?
4. **COMMUNICATE** Draw a picture that shows the many ways that people use water and air.
5. **CRITICAL THINKING** *Evaluate* Wood is a fuel. Do you think wood is a renewable or nonrenewable resource? Explain.

WHY IT MATTERS THINK ABOUT IT
Describe some of the places you go each day. What do you do at these places? How do you get there?

WHY IT MATTERS WRITE ABOUT IT
Gasoline is a fuel that powers cars, buses, and trucks. How would your life be different if there were no gasoline?

READING SKILL How does coal form? Draw a diagram that shows the sequence of events.

SCIENCE MAGAZINE

POSITIVELY PLASTIC

What comes in every color of the rainbow and can be found in almost every home in America? Something that's made of plastic!

We use plastic wrap to protect our foods. We put our garbage in plastic bags or plastic cans. We sit on plastic chairs, play with plastic toys, drink from plastic cups, and wash our hair with shampoo from plastic bottles! We know how useful it is, but exactly what is it?

Plastic doesn't grow in nature. It's made by mixing certain things together. We call it a produced or manufactured material. Plastic was first made in the 1860s from plants, such as wood and cotton. That plastic was soft and burned easily.

The first modern plastics were made in the 1930s. Today plastics are made from oil and natural gas. It's true! Most clear plastic starts out as thick, black oil. That plastic coating inside a pan begins as natural gas.

Over the years, hundreds of different plastics have been

Science, Technology, and Society

developed. Some are hard and strong. Some are soft and bendable. Some are clear. Some are many-colored. There's a plastic for almost every need. Scientists continue to experiment with plastics. They hope to find even more ways to use them!

DISCUSSION STARTER

1. Most plastics don't rot. What's good about that?
2. What problems occur because plastics don't rot?

To learn more about plastics, visit *www.mhschool.com/science* and enter the keyword MADE.

*inter*NET CONNECTION

Topic 6
EARTH SCIENCE

WHY IT MATTERS

Conserving resources means making sure there are enough of them for the future.

SCIENCE WORDS

pollution what happens when harmful substances get into water, air, or land

reduce to make less of something

reuse to use something again

recycle to treat something so it can be used again

Conserving Earth's Resources

What has made the water dirty in this picture? Living things need clean water. People use water for drinking, cleaning, and cooking. Plants need clean water in order to grow. Dirty water can cause sickness and even death in animals and people. Do you think this water is safe to use?

EXPLORE

HYPOTHESIZE One way we use water is to keep ourselves clean. When you wash your hands, where do the dirt and grime go? Do they stay in the water? Write a hypothesis in your *Science Journal*. How might you test your ideas?

EXPLORE ACTIVITY

Investigate What Happens When Materials Get into Water

Explore what happens when materials are put into water.

MATERIALS
- small jar with lid
- water
- oil
- plastic spoon
- liquid soap
- *Science Journal*

PROCEDURES

1. **OBSERVE** Pour some water into your jar and add a spoonful of the oil. How does the water look? Where is the oil? Record your observations in your *Science Journal*.

2. **OBSERVE** Shake the jar. Set the jar down and watch it for a few minutes. What happens? Can you scoop the oil out with your spoon?

3. **EXPERIMENT** Add a few drops of liquid soap to the jar and shake the jar again. What happened to the oil? Can you clean the oil out now?

CONCLUDE AND APPLY

1. **DESCRIBE** What happened when you shook the jar with the oil and water in it?

2. **IDENTIFY** Where did the oil and soap go when you shook the jar a second time?

GOING FURTHER: Apply

3. **INFER** Where do you think dirt and grime go when you wash a car or dirty dishes?

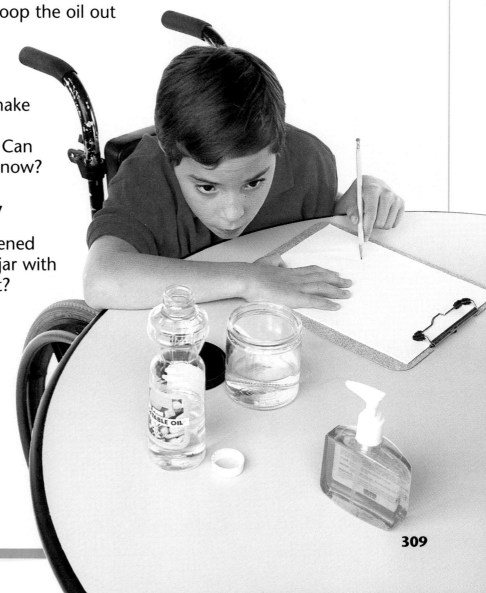

309

What Happens When Materials Get into Water?

As we use Earth's water, air, and land, we add materials to them. For example, when you wash your hands, you rinse soap and dirt into the water. Water is cleaned naturally by flowing through layers of rock in the ground.

The Explore Activity shows that a problem may develop when too many materials get into the water. This problem is **pollution** (pə lü′shən). Pollution occurs when harmful substances get into the water, air, or land. When these resources become polluted they become unsafe to use.

Some pollution happens naturally. Polluted water happens naturally when too much sand or soil settles in a lake. Then the lake isn't a good place for fish and plants to live. Volcanic eruptions and forest fires pollute the air with dust, gas, and ashes.

Forest fire

Human actions cause pollution, too. People add things like soap and *fertilizers* (fur′tə lī′zər) to water. A fertilizer is used to help plants grow. Fertilizers can soak through the ground and into water supplies.

Cleaning Water

HYPOTHESIZE Can you clean water by letting it flow through rocks? Write a hypothesis in your *Science Journal*.

PROCEDURES

1. **MODEL** Place the funnel inside the bottom half of the plastic bottle. Put a layer of gravel in the funnel and cover it with a layer of sand.

2. **MODEL** Mix a cup of water with a little soil and some crushed leaves. Draw a picture of the water in your *Science Journal*. Slowly pour the mixture into the funnel.

CONCLUDE AND APPLY

1. **DESCRIBE** How does the water look when it filters through to the bottom half of the bottle?

2. **INFER** Where is the soil?

MATERIALS
- plastic funnel
- bottom half of plastic bottle
- gravel
- sand
- measuring cup
- water
- spoonful of soil
- dead leaf
- cup
- *Science Journal*

311

Smog hangs over the city of Los Angeles. Smog is fog that has been polluted by smoke.

Mario Molina studies the effects of chemicals called *fluorocarbons* on Earth's atmosphere. The results of his research have been used to create laws that help to control pollution.

What Other Materials Pollute Earth's Land and Air?

Cars, airplanes, and factories add gases and chemicals to the air. Rain carries materials in the air to Earth. This polluted rain water flows into rivers, lakes, and oceans, and soaks into the ground.

In some places, land is polluted by enormous amounts of trash. As Earth's population grows, more trash is produced.

There are many people and other living things on Earth. All living things need to use Earth's resources. There are several things people can do to help conserve the resources that all living things need.

One way for people to conserve resources is to **reduce** (ri düs′) the amount of things they use. To reduce something is to make less of it. You can reduce trash by buying products that come in less packaging. You can use both sides of notebook paper before throwing it out. You can reduce the amount of water you use by turning off the faucet while you brush your teeth.

ONE PERSON'S TRASH EACH YEAR

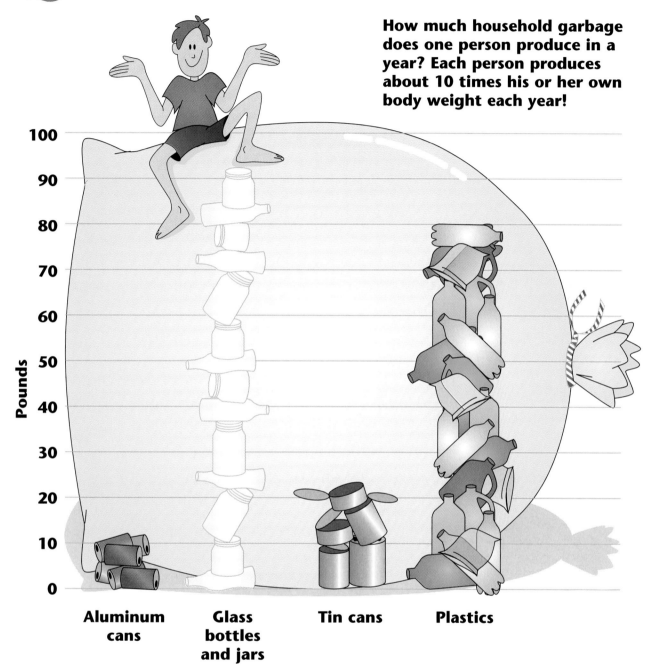

How much household garbage does one person produce in a year? Each person produces about 10 times his or her own body weight each year!

READING CHARTS

1. **WRITE** How many pounds of cans, bottles, and jars does one person use each year?
2. **DISCUSS** How can you figure out about how much garbage you produce each year? You'll have to do some multiplication to find out.

What Other Things Can You Do to Conserve Resources?

Another way for people to conserve resources is to **reuse** (rē ūs′) things. To reuse something is to use it again. You can make an old sock into a puppet for your younger brother's birthday. Or you can reuse plastic grocery bags as trash bags. Plastic containers can be reused to store leftovers, to pack lunches, and to hold small items such as paper clips and crayons.

A third way to conserve Earth's resources is to **recycle** (rē sī′kəl). To recycle something means to treat it so that it can be used again. Glass, plastic, paper, and aluminum and tin cans are examples of things that can be recycled. Paper can be recycled, too.

Different communities have different rules about recycling. These rules may tell you what can be recycled in your community. They may also tell you where to take items to be recycled. You can even buy things that have been made from recycled materials. When you do this, you are conserving resources by reusing them!

The numbers inside the recycling symbol on a plastic container are codes. The number tells the recycling plant what type of material the container is made from.

Take a survey of plastics in your house. Can you find an example of each type?

Brain Power

How could you reduce the amount of trash thrown away in your classroom?

WHY IT MATTERS

As the number of people in the world increases, the need for resources also increases. People can help conserve resources. When you work to conserve resources, you are working to make sure there are enough of them for the future.

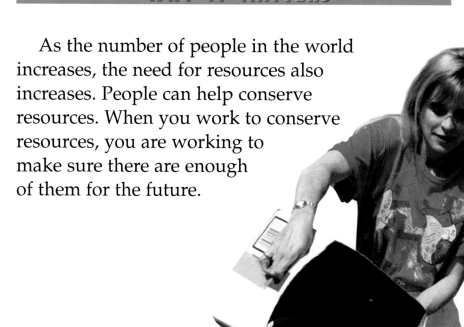

This person is recycling.

REVIEW

1. What happens when too many materials get into water?
2. What is pollution?
3. How can water, air, and land become polluted?
4. **IDENTIFY** Name three ways to reuse a plastic container.
5. **CRITICAL THINKING** *Apply* You are given the job of Resource Monitor in your classroom. Write a description of what you would do to make sure the class resources are used wisely.

WHY IT MATTERS THINK ABOUT IT
Estimate the amount of trash you throw away each day. What kinds of things do you throw away?

WHY IT MATTERS WRITE ABOUT IT
What can you do to reduce the amount of trash you throw away? What can you reuse? What can you recycle?

SCIENCE MAGAZINE

Kids Did It!

Pollution at Paint Creek

Is Paint Creek polluted? Nine kids who lived near the creek wanted to find out. The kids, ages six to ten, were part of a group in a Ypsilanti (ip'sə lan'tē), Michigan, summer program.

Kids in the program chose several tests to help answer their question.

The kids created data sheets to record their findings. Was the creek polluted? After checking all the data, the kids didn't agree on an answer. They did agree that they learned a lot about pollution and about working together! They also agreed it was fun to have their story and picture in the local newspaper!

DISCUSSION STARTER

1. What tests did the group decide to use?
2. What did the group learn from the project?

Test 1
Water temperature
Healthy water in a stream like Paint Creek should be cool, about 17°C or 63°F.

Test 2
Water speed
Water moves faster in a healthy stream.

Test 3
Number of insects
The more insects, the better. They're fish food!

Test 4
Water quality
How the water looks, smells, and feels. The group decided they better not taste it!

To learn more about checking water pollution, visit **www.mhschool.com/science** and enter the keyword CREEK.

interNET CONNECTION

CHAPTER 10 REVIEW

SCIENCE WORDS

natural resource p.292
nonrenewable resource p.302
pollution p.310
recycle p.314
reduce p.312
renewable resource p.296
reuse p.314

USING SCIENCE WORDS

Number a paper from 1 to 10. Fill in 1 to 5 with words from the list above.

1. A material on Earth that is necessary or useful to people is a __?__.
2. A resource that cannot be reused or replaced in a useful amount of time is a __?__.
3. When too many materials get into water, there is __?__.
4. When you __?__ something, you find a way to use it again.
5. A resource that can be used over and over again is a __?__.

6–10. Pick five words from the list above. Include all words that were not used in 1 to 5. Write each word in a sentence.

UNDERSTANDING SCIENCE IDEAS

11. How does water become polluted?
12. If polluted water can be made clean again, why is it important not to pollute it?

USING IDEAS AND SKILLS

13. **READING SKILL: SEQUENCE OF EVENTS** How is soil formed?
14. **MEASURE** Look at the graduated cylinder. How much water does it hold?

15. **THINKING LIKE A SCIENTIST** You want to know if a lake near your school is polluted. How might you test the lake to find out?

PROBLEMS and PUZZLES

Growth Spurt How is soil different from sand? Experiment. Fill a can with soil and a second can with sand. Plant bean seeds in each can. What happens after one week? Two weeks? Why did you get these results?

UNIT 5 REVIEW

SCIENCE WORDS

earthquake p.282
erosion p.272
glacier p.272
hurricane p.280
mineral p.260
natural resource p.292
nonrenewable resource p.302
plain p.264
pollution p.310
recycle p.314
reduce p.312
renewable resource p.296
reuse p.314
valley p.264
volcano p.283
weathering p.270

USING SCIENCE WORDS

Number a paper from 1 to 10. Beside each number write the word or words that best completes the sentence.

1. Rocks are made up of __?__.

2. The landform that makes up the low area between hills is a __?__.

3. The wearing away of rock by ice, water, or wind is __?__.

4. A huge mass of ice and snow that moves is called a(n) __?__.

5. A sudden movement of Earth's crust that can cause the ground to shake is a(n) __?__.

6. Lava pours out of the hole of a __?__.

7. Materials from Earth that are useful or necessary are __?__.

8. Coal is a __?__ because it cannot be reused or replaced in a useful amount of time.

9. Smoke from a fire can cause air __?__.

10. You can __?__ plastic bags so that the plastic can be used again.

UNDERSTANDING SCIENCE IDEAS

Write 11 to 15. For each number write the letter for the best answer. You may wish to use the hints provided.

11. What tells you that a rock is soft?
    a. The rock scratches easily.
    b. The rock is very heavy.
    c. The rock is gray and white.
    d. The rock is smooth.
    (Hint: Read page 261.)

12. Weathering makes rocks
    a. larger
    b. break apart
    c. melt
    d. harder
    (Hint: Read pages 270–272.)

13. Hurricanes often cause
    a. earthquakes
    b. rocks to break
    c. soil to be washed away
    d. caves to form
    (Hint: Read pages 279–280.)

14. Natural gas and oil are
    a. renewable resources
    b. minerals
    c. nonrenewable resources
    d. formed by glaciers
    (Hint: Read page 302.)

15. Treating old materials to make them into new products is
    a. recycling
    b. pollution
    c. weathering
    d. reusing
    (Hint: Read pages 313–314.)

UNIT 5 REVIEW

USING IDEAS AND SKILLS

16. In what ways can rocks be different from each other? Give three examples.

17. Name two renewable resources and describe how people use them.

18. **MEASURE** Your kitchen sink has a leaky faucet. How can you find out how much water is wasted each day?

19. Describe two ways that an earthquake might be dangerous to people.

THINKING LIKE A SCIENTIST

20. **HYPOTHESIZE** You visit a lake that was once clear and full of fish. The lake water now looks dirty and the fish are all gone. What do you think happened?

WRITING IN YOUR JOURNAL

SCIENCE IN YOUR LIFE
List some things that people recycle. Explain how you might make recycling easier for people to do.

PRODUCT ADS
The labels of some laundry soaps tell you that the soaps break down into harmless materials. Why is this a good characteristic for something that will be put into water?

HOW SCIENTISTS WORK
In this unit you have learned that weathering and erosion can change landforms. Scientists use their skills of observation and measuring to study these changes. Why do you think it is important to understand how landforms change?

Design your own Experiment

Would clay or sand be washed away more easily by rain? Design an experiment to find out. Review your experiment with your teacher before trying it out.

For help in reviewing this unit, visit **www.mhschool.com/science**

UNIT 5 REVIEW

PROBLEMS and PUZZLES

Be a Rockhound at Home

You don't have to go outside to start a rock and mineral collection. You can find rocks and minerals in your house or apartment. Here are some examples to get you started:

table salt (halite)

pencil lead (graphite)

pumice (volcanic rock)

You can store and display your collection in egg cartons or in reusable plastic bags and containers.

The Garbage Problem

Food garbage is a part of the things people throw away. Is there a way this garbage can be reused or recycled? Find out!

Line a shoebox with plastic. Poke several holes through the box and the plastic on each side. Fill the box about half full with crushed dead leaves. Add two handfuls of soil and 1/2 cup of water. Do this again, then mix the materials with your hands. The mixture should be damp, but not wet. Add more soil or water if you need to.

Next mix in two or three tablespoons of food garbage—vegetable peelings, eggshells, or coffee grounds. Don't include meat scraps, milk, or cheese! Put about 150 red worms in the box and cover it with the lid.

For best results, you will need to keep this project going for at least 10 days. Check the box each day. Add water and soil if needed. Add more kitchen scraps as they disappear. What changes take place?

A Dune on the Move

Dunes are mounds of sand made by the wind. The wind picks up the sand from the side of the dune that is not steep. Then it leaves the sand on the steeper side. See if you can put the pictures in the correct order.

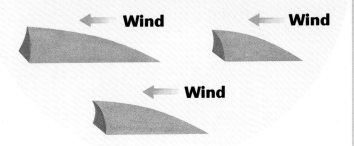

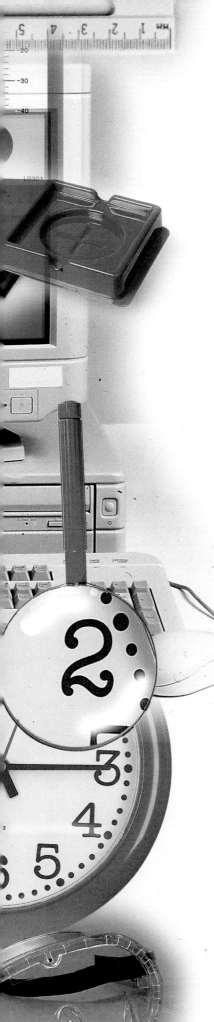

REFERENCE SECTION

HANDBOOK

MEASUREMENTS R2
SAFETY R4
COLLECT DATA
- HAND LENS R6
- MICROSCOPE R7
- COMPASS R8
- TELESCOPE R9
- CAMERA, TAPE RECORDER, MAP, AND COMPASS ... R10

MAKE MEASUREMENTS
- LENGTH R11
- TIME R12
- VOLUME R13
- MASS R14
- WEIGHT/FORCE R16
- TEMPERATURE R17

MAKE OBSERVATIONS
- WEATHER R18
- SYSTEMS R19

REPRESENT DATA
- GRAPHS R20
- MAPS R22
- TABLES AND CHARTS R23

USE TECHNOLOGY
- COMPUTER R24
- CALCULATOR R26

GLOSSARY R27

INDEX R39

MEASUREMENTS

Temperature

1. The temperature is 77 degrees Fahrenheit.
2. That is the same as 25 degrees Celsius.
3. Water boils at 212 degrees Fahrenheit.
4. Water freezes at 0 degrees Celsius.

Length and Area

1. This classroom is 10 meters wide and 20 meters long.
2. That means the area is 200 square meters.

Mass and Weight

1. That baseball bat weighs 32 ounces.
2. 32 ounces is the same as 2 pounds.
3. The mass of the bat is 907 grams.

R2

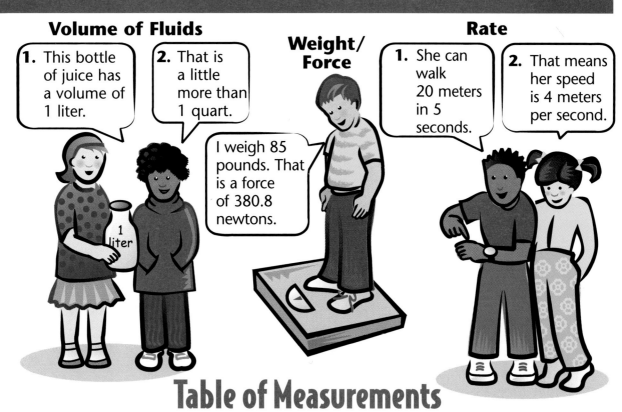

Table of Measurements

SI (International System) of Units

Temperature
Water freezes at 0 degrees Celsius (°C) and boils at 100°C.

Length and Distance
10 millimeters (mm) = 1 centimeter (cm)
100 centimeters = 1 meter (m)
1,000 meters = 1 kilometer (km)

Volume
1 cubic centimeter (cm^3) = 1 milliliter (mL)
1,000 milliliters = 1 liter (L)

Mass
1,000 milligrams (mg) = 1 gram (g)
1,000 grams = 1 kilogram (kg)

Area
1 square kilometer (km^2) = 1 km x 1 km
1 hectare = 10,000 square meters (m^2)

Rate
m/s = meters per second
km/h = kilometers per hour

Force
1 newton (N) = 1 kg x m/s^2

English System of Units

Temperature
Water freezes at 32 degrees Fahrenheit (°F) and boils at 212°F.

Length and Distance
12 inches (in.) = 1 foot (ft)
3 feet = 1 yard (yd)
5,280 feet = 1 mile (mi)

Volume of Fluids
8 fluid ounces (fl oz) = 1 cup (c)
2 cups = 1 pint (pt)
2 pints = 1 quart (qt)
4 quarts = 1 gallon (gal)

Weight
16 ounces (oz) = 1 pound (lb)
2,000 pounds = 1 ton (T)

Rate
mph = miles per hour

SAFETY

In the Classroom

The most important part of doing any experiment is doing it safely. You can be safe by paying attention to your teacher and doing your work carefully. Here are some other ways to stay safe while you do experiments.

Before the Experiment

- Read all of the directions. Make sure you understand them. When you see ▰, be sure to follow the safety rule.
- Listen to your teacher for special safety directions. If you don't understand something, ask for help.

During the Experiment

- Wear safety goggles when your teacher tells you to wear them and whenever you see ⌒.
- Wear a safety apron if you work with anything messy or anything that might spill.
- If you spill something, wipe it up right away or ask your teacher for help.
- Tell your teacher if something breaks. If glass breaks do not clean it up yourself.
- Keep your hair and clothes away from open flames. Tie back long hair and roll up long sleeves.
- Be careful around a hot plate. Know when it is on and when it is off. Remember that the plate stays hot for a few minutes after you turn it off.
- Keep your hands dry around electrical equipment.
- Don't eat or drink anything during the experiment.

After the Experiment

- Put equipment back the way your teacher tells you.
- Dispose of things the way your teacher tells you.
- Clean up your work area and wash your hands.

In the Field

- Always be accompanied by a trusted adult—like your teacher or a parent or guardian.
- Never touch animals or plants without the adult's approval. The animal might bite. The plant might be poison ivy or another dangerous plant.

Responsibility

Acting safely is one way to be responsible. You can also be responsible by treating animals, the environment, and each other with respect in the class and in the field.

Treat Living Things with Respect

- If you have animals in the classroom, keep their homes clean. Change the water in fish tanks and clean out cages.
- Feed classroom animals the right amounts of food.
- Give your classroom animals enough space.
- When you observe animals, don't hurt them or disturb their homes.
- Find a way to care for animals while school is on vacation.

Treat the Environment with Respect

- Do not pick flowers.
- Do not litter, including gum and food.
- If you see litter, ask your teacher if you can pick it up.
- Recycle materials used in experiments. Ask your teacher what materials can be recycled instead of thrown away. These might include plastics, aluminum, and newspapers.

Treat Each Other with Respect

- Use materials carefully around others so that people don't get hurt or get stains on their clothes.
- Be careful not to bump people when they are doing experiments. Do not disturb or damage their experiments.
- If you see that people are having trouble with an experiment, help them.

COLLECT DATA

Use a Hand Lens

You use a hand lens to magnify an object, or make the object look larger. With a hand lens, you can see details that would be hard to see without the hand lens.

Magnify a Piece of Cereal

1. Place a piece of your favorite cereal on a flat surface. Look at the cereal carefully. Draw a picture of it.
2. Hold the hand lens so that it is just above the cereal. Look through the lens, and slowly move it away from the cereal. The cereal will look larger.
3. Keep moving the hand lens until the cereal begins to look blurry. Then move the lens a little closer to the cereal until you can see it clearly.
4. Draw a picture of the cereal as you see it through the hand lens. Fill in details that you did not see before.
5. Repeat this activity using objects you are studying in science. It might be a rock, some soil, a flower, a seed, or something else.

Use a Microscope

Hand lenses make objects look several times larger. A microscope, however, can magnify an object to look hundreds of times larger.

Examine Salt Grains

1. Place the microscope on a flat surface. Always carry a microscope with both hands. Hold the arm with one hand, and put your other hand beneath the base.
2. Look at the drawing to learn the different parts of the microscope.
3. Move the mirror so that it reflects light up toward the stage. Never point the mirror directly at the Sun or a bright light.
4. Place a few grains of salt on the slide. Put the slide under the stage clips on the stage. Be sure that the salt grains are over the hole in the stage.
5. Look through the eyepiece. Turn the focusing knob slowly until the salt grains come into focus.
6. Draw what the grains look like through the microscope.
7. Look at other objects through the microscope. Try a piece of leaf, a strand of human hair, or a pencil mark.

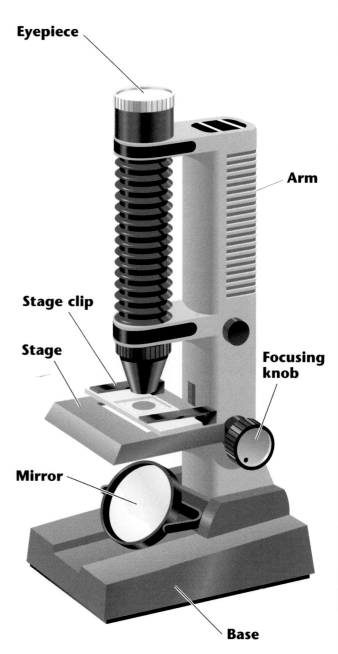

COLLECT DATA

Use a Compass

You use a compass to find directions. A compass is a small, thin magnet that swings freely, like a spinner in a board game. One end of the magnet always points north. This end is the magnet's north pole. How does a compass work?

1. Place the compass on a surface that is not made of magnetic material. A wooden table or a sidewalk works well.
2. Find the magnet's north pole. The north pole is marked in some way, usually with a color or an arrowhead.
3. Notice the letters N, E, S, and W on the compass. These letters stand for the directions north, east, south, and west. When the magnet stops swinging, turn the compass so that the N lines up with the north pole of the magnet.
4. Face to the north. Then face to the east, to the south, and to the west.
5. Repeat this activity by holding the compass in your hand and then at different places indoors and outdoors.

Use a Telescope

A telescope makes faraway objects, like the Moon, look larger. A telescope also lets you see stars that are too faint to see with just your eyes.

Look at the Moon

1. Look at the Moon in the night sky. Draw a picture of what you see. Draw as many details as you can.
2. Point a telescope toward the Moon. Look through the eyepiece of the telescope. Move the telescope until you see the Moon. Turn the knob until the Moon comes into focus.
3. Draw a picture of what you see. Include details. Compare your two pictures.

Look at the Stars

1. Find the brightest star in the sky. Notice if there are any other stars near it.
2. Point a telescope toward the brightest star. Look through the eyepiece and turn the knob until the stars come into focus. Move the telescope until you find the brightest star.
3. Can you see stars through the telescope that you cannot see with just your eyes?

COLLECT DATA

Use a Camera, Tape Recorder, Map, and Compass

Camera

You can use a camera to record what you observe in nature. Keep these tips in mind.

1. Hold the camera steady. Gently press the button so that you do not jerk the camera.
2. Try to take pictures with the Sun at your back. Then your pictures will be bright and clear.
3. Don't get too close to the subject. Without a special lens, the picture could turn out blurry.
4. Be patient. If you are taking a picture of an animal, you may have to wait for the animal to appear.

Tape Recorder

You can record observations on a tape recorder. This is sometimes better than writing notes because a tape recorder can record your observations at the exact time you are making them. Later you can listen to the tape and write down your observations.

Map and Compass

When you are busy observing nature, it might be easy to get lost. You can use a map of the area and a compass to find your way. Here are some tips.

1. Lightly mark on the map your starting place. It might be the place where the bus parked.
2. Always know where you are on the map compared to your starting place. Watch for landmarks on the map, such as a river, a pond, trails, or buildings.
3. Use the map and compass to find special places to observe, such as a pond. Look at the map to see which direction the place is from you. Hold the compass to see where that direction is.
4. Use your map and compass with a friend.

MAKE MEASUREMENTS

Length

Find Length with a Ruler

1. Look at this section of a ruler. Each centimeter is divided into 10 millimeters. How long is the paper clip?
2. The length of the paper clip is 3 centimeters plus 2 millimeters. You can write this length as 3.2 centimeters.
3. Place the ruler on your desk. Lay a pencil against the ruler so that one end of the pencil lines up with the left edge of the ruler. Record the length of the pencil.
4. Trade your pencil with a classmate. Measure and record the length of each other's pencils. Compare your answers.

Measuring Area

Area is the amount of surface something covers. To find the area of a rectangle, multiply the rectangle's length by its width. For example, the rectangle here is 3 centimeters long and 2 centimeters wide. Its area is 3 cm x 2 cm = 6 square centimeters. You write the area as 6 cm^2.

1. Find the area of your science book. Measure the book's length to the nearest centimeter. Measure its width.
2. Multiply the book's length by its width. Remember to put the answer in cm^2.

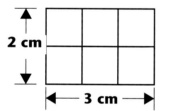

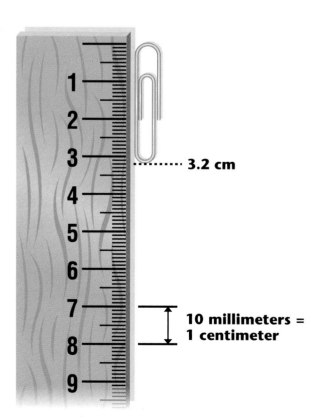

MAKE MEASUREMENTS

Time

You use timing devices to measure how long something takes to happen. Some timing devices you use in science are a clock with a second hand and a stopwatch. Which one is more accurate?

Comparing a Clock and a Stopwatch

1. Look at a clock with a second hand. The second hand is the hand that you can see moving. It measures seconds.
2. Get an egg timer with falling sand or some device like a windup toy that runs down after a certain length of time. When the second hand of the clock points to 12, tell your partner to start the egg timer. Watch the clock while the sand in the egg timer is falling.
3. When the sand stops falling, count how many seconds it took. Record this measurement. Repeat the activity, and compare the two measurements.
4. Switch roles with your partner.
5. Look at a stopwatch. Click the button on the top right. This starts the time. Click the button again. This stops the time. Click the button on the top left. This sets the stopwatch back to zero. Notice that the stopwatch tells time in hours, minutes, seconds, and hundredths of a second.
6. Repeat the activity in steps 1–3, but use the stopwatch instead of a clock. Make sure the stopwatch is set to zero. Click the top right button to start timing.

Click the button again when the sand stops falling. Make sure you and your partner time the sand twice.

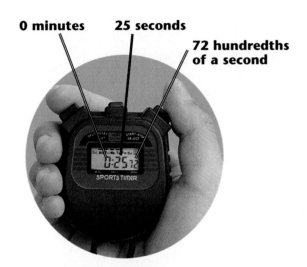

0 minutes   25 seconds
72 hundredths of a second

More About Time

1. Use the stopwatch to time how long it takes an ice cube to melt under cold running water. How long does an ice cube take to melt under warm running water?
2. Match each of these times with the action you think took that amount of time.

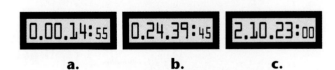

a.   b.   c.

1. A Little League baseball game
2. Saying the Pledge of Allegiance
3. Recess

R12

Volume

Have you ever used a measuring cup? Measuring cups measure the volume of liquids. Volume is the amount of space something takes up. To bake a cake, you might measure the volume of water, vegetable oil, or melted butter. In science you use special measuring cups called beakers and graduated cylinders. These containers are marked in milliliters (mL).

Measure the Volume of a Liquid

1. Look at the beaker and at the graduated cylinder. The beaker has marks for each 25 mL up to 200 mL. The graduated cylinder has marks for each 1 mL up to 100 mL.

2. The surface of the water in the graduated cylinder curves up at the sides. You measure the volume by reading the height of the water at the flat part. What is the volume of water in the graduated cylinder? How much water is in the beaker? They both contain 75 mL of water.

3. Pour 50 mL of water from a pitcher into a graduated cylinder. The water should be at the 50-mL mark on the graduated cylinder. If you go over the mark, pour a little water back into the pitcher.

4. Pour the 50 mL of water into a beaker.

5. Repeat steps 3 and 4 using 30 mL, 45 mL, and 25 mL of water.

6. Measure the volume of water you have in the beaker. Do you have about the same amount of water as your classmates?

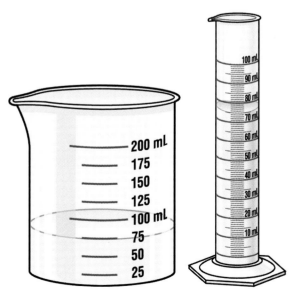

MAKE MEASUREMENTS

Mass

Mass is the amount of matter an object has. You use a balance to measure mass. To find the mass of an object, you balance it with objects whose masses you know. Let's find the mass of a box of crayons.

Measure the Mass of a Box of Crayons

1. Place the balance on a flat, level surface. Check that the two pans are empty and clean.
2. Make sure the empty pans are balanced with each other. The pointer should point to the middle mark. If it does not, move the slider a little to the right or left to balance the pans.
3. Gently place a box of crayons on the left pan. The pan will drop lower.
4. Add masses to the right pan until the pans are balanced. You can use paper clips.
5. Count the number of paper clips that are in the right pan. Two paper clips equal about one gram. What is the mass of the box of crayons? Record the number. After the number, write a *g* for "grams."

Predict the Mass of More Crayons

1. Leave the box of crayons and the masses on the balance.
2. Get two more crayons. If you put them in the pan with the box of crayons, what do you think the mass of all the crayons will be? Write down what you predict the total mass will be.
3. Check your prediction. Gently place the two crayons in the left pan. Add masses, such as paper clips, to the right pan until the pans are balanced.
4. Calculate the mass as you did before. Record this number. How close is it to your prediction?

More About Mass

What was the mass of all your crayons? It was probably less than 100 grams. What would happen if you replaced the crayons with a pineapple? You may not have enough masses to balance the pineapple. It has a mass of about 1,000 grams. That's the same as 1 kilogram because *kilo* means "1,000."

MAKE MEASUREMENTS

Weight/Force

You use a spring scale to measure weight. An object has weight because the force of gravity pulls down on the object. Therefore, weight is a force. Like all forces weight is measured in newtons (N).

Measure the Weight of an Object

1. Look at your spring scale to see how many newtons it measures. See how the measurements are divided. The spring scale shown here measures up to 10 N. It has a mark for every 1 N.
2. Hold the spring scale by the top loop. Put the object to be measured on the bottom hook. If the object will not stay on the hook, place it in a net bag. Then hang the bag from the hook.
3. Let go of the object slowly. It will pull down on a spring inside the scale. The spring is connected to a pointer. The pointer on the spring scale shown here is a small bar.
4. Wait for the pointer to stop moving. Read the number of newtons next to the pointer. This is the object's weight. The mug in the picture weighs 3 N.

More About Spring Scales

You probably weigh yourself by standing on a bathroom scale. This is a spring scale. The force of your body stretches or presses a spring inside the scale. The dial on the scale is probably marked in pounds—the English unit of weight. One pound is equal to about 4.5 newtons.

Here are some spring scales you may have seen.

Temperature

Temperature is how hot or cold something is. You use a thermometer to measure temperature. A thermometer is made of a thin tube with colored liquid inside. When the liquid gets warmer, it expands and moves up the tube. When the liquid gets cooler, it contracts and moves down the tube. You may have seen most temperatures measured in degrees Fahrenheit (°F). Scientists measure temperature in degrees Celsius (°C).

Read a Thermometer

1. Look at the thermometer shown here. It has two scales—a Fahrenheit scale and a Celsius scale. Every 20 degrees on each scale has a number.
2. What is the temperature shown on the thermometer? At what temperature does water freeze? Give your answers in °F and in °C.

How Is Temperature Measured?

1. Fill a large beaker about one-half full of cool water. Find the temperature of the water by holding a thermometer in the water. Do not let the bulb at the bottom of the thermometer touch the sides or bottom of the beaker.
2. Keep the thermometer in the water until the liquid in the tube stops moving—about a minute. Read and record the temperature on the Celsius scale.
3. Fill another large beaker one-half full of warm water from a faucet. Be careful not to burn yourself by using hot water.
4. Find and record the temperature of the warm water just as you did in steps 1 and 2.

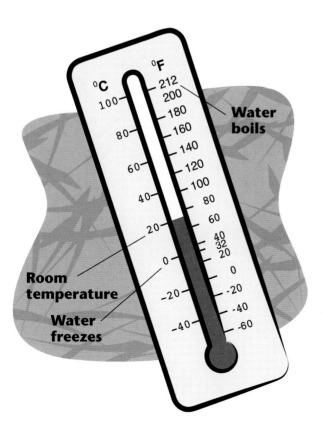

R17

MAKE OBSERVATIONS

Weather

What was the weather like yesterday? What is it like today? The weather changes from day to day. You can observe different parts of the weather to find out how it changes.

Measure Temperature

1. Use a thermometer to find the air temperature outside. Look at page R17 to review thermometers.
2. Hold a thermometer outside for two minutes. Then read and record the temperature.
3. Take the temperature at the same time each day for a week. Record it in a chart.

Observe Wind Speed and Direction

1. Observe how the wind is affecting things around you. Look at a flag or the branches of a tree. How hard is the wind blowing the flag or branches? Observe for about five minutes. Write down your observations.
2. Hold a compass to see which direction the wind is coming from. Write down this direction.
3. Observe the wind each day for a week. Record your observations in your chart.

Observe Clouds, Rain, and Snow

1. Observe how much of the sky is covered by clouds. Use these symbols to record the cloud cover in your chart each day.

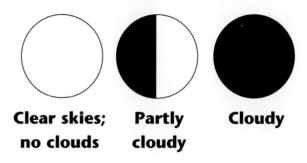

Clear skies; no clouds   Partly cloudy   Cloudy

2. Record in your chart if it is raining or snowing.
3. At the end of the week, how has the weather changed from day to day?

R18

Systems

What do a toy car, a tomato plant, and a yo-yo have in common? They are all systems. A system is a set of parts that work together to form a whole. Look at the three systems below. Think of how each part helps the system work.

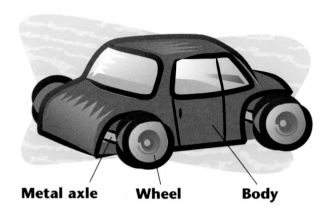

Metal axle **Wheel** **Body**

This system has three main parts—the body, the axles, and the wheels. Would the system work well if the axles could not turn?

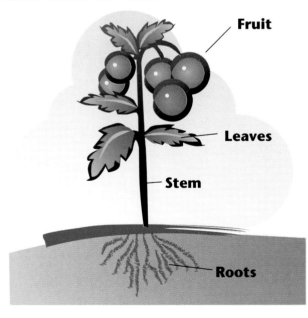

Fruit
Leaves
Stem
Roots

In this system roots take in water, and leaves make food. The stem carries water and food to different parts of the plant. What would happen if you cut off all the leaves?

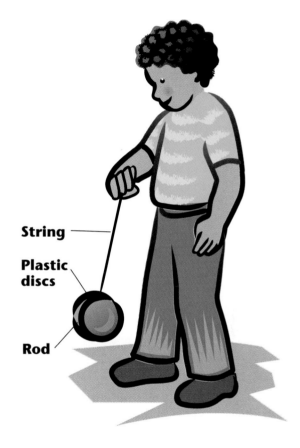

String
Plastic discs
Rod

Even simple things can be systems. How do all the parts of the yo-yo work together to make the toy go up and down?

Look for some other systems at school, at home, and outside. Remember to look for things that are made of parts. List the parts. Then describe how you think each part helps the system work.

R19

REPRESENT DATA

Make Graphs to Organize Data

When you do an experiment in science, you collect information. To find out what your information means, you can organize it into graphs. There are many kinds of graphs.

Bar Graphs

A bar graph uses bars to show information. For example, suppose you are growing a plant. Every week you measure how high the plant has grown. Here is what you find.

Week	Height (cm)
1	1
2	3
3	6
4	10
5	17
6	20
7	22
8	23

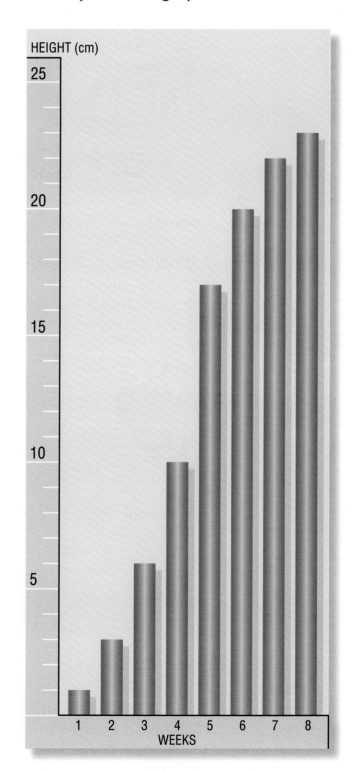

The bar graph at right organizes the measurements you collected so that you can easily compare them.

1. Look at the bar for week 2. Put your finger at the top of the bar. Move your finger straight over to the left to find how many centimeters the plant grew by the end of week 2.
2. Between which two weeks did the plant grow most?
3. When did plant growth begin to level off?

R20

Pictographs

A pictograph uses symbols, or pictures, to show information. Suppose you collect information about how much water your family uses each day. Here is what you find.

Activity	Water Used Each Day (L)
Drinking	10
Showering	180
Bathing	240
Brushing teeth	80
Washing dishes	140
Washing hands	30
Washing clothes	280
Flushing toilet	90

You can organize this information into the pictograph shown here. The pictograph has to explain what the symbol on the graph means. In this case, each bottle means 20 liters of water. A half bottle means half of 20, or 10 liters of water.

1. Which activity uses the most water?
2. Which activity uses the least water?

Make a Graph

Suppose you do an experiment to find out how far a toy car rolls on different surfaces. The results of your experiment are shown below.

Surface	Distance Car Rolled (cm)
Wood Floor	525
Sidewalk	325
Carpet Floor	150
Tile Floor	560
Grass	55

1. Decide what kind of graph would best show these results.
2. Make your graph.

A Family's Daily Use of Water

REPRESENT DATA
Make Maps to Show Information

Locate Places

A map is a drawing that shows an area from above. Most maps have numbers and letters along the top and side. They help you find places easily. For example, what if you wanted to find the library on the map below. It is located at D7. Place a finger on the letter D along the side of the map and another finger on the number 7 at the top. Then move your fingers straight across and down the map until they meet. The library is located where D and 7 meet, or very nearby.

1. What building is located at G3?
2. The hospital is located three blocks south and three blocks east of the library. What is its number and letter?
3. Make a map of an area in your community. It might be a park or the area between your home and school. Include numbers and letters along the top and side. Use a compass to find north, and mark north on your map. Exchange maps with classmates.

Idea Maps

The map below left shows how places are connected to each other. Idea maps, on the other hand, show how ideas are connected to each other. Idea maps help you organize information about a topic.

Look at the idea map below. It connects ideas about water. This map shows that Earth's water is either fresh water or salt water. The map also shows four sources of fresh water. You can see that there is no connection between "rivers" and "salt water" on the map. This reminds you that salt water does not flow in rivers.

Make an idea map about a topic you are learning in science. Your map can include words, phrases, or even sentences. Arrange your map in a way that makes sense to you and helps you understand the ideas.

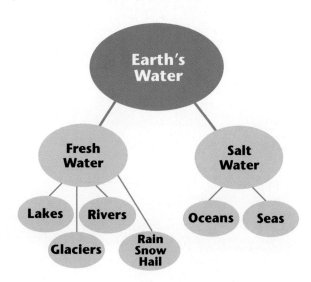

Make Tables and Charts to Organize Data

Tables help you organize data during experiments. Most tables have columns that run up and down, and rows that run across. The columns and rows have headings that tell you what kind of data goes in each part of the table.

A Sample Table

What if you are going to do an experiment to find out how long different kinds of seeds take to sprout? Before you begin the experiment, you should set up your table. Follow these steps.

1. In this experiment you will plant 20 radish seeds, 20 bean seeds, and 20 corn seeds. Your table must show how many of each kind of seed sprouted on days 1, 2, 3, 4, and 5.
2. Make your table with columns, rows, and headings. You might use a computer. Some computer programs let you build a table with just the click of a mouse. You can delete or add columns and rows if you need to.
3. Give your table a title. Your table could look like the one here.

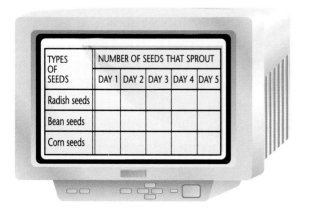

Make a Table

Now what if you are going to do an experiment to find out how temperature affects the sprouting of seeds? You will plant 20 bean seeds in each of two trays. You will keep each tray at a different temperature, as shown below, and observe the trays for seven days. Make a table that you can use for this experiment.

Make a Chart

A chart is simply a table with pictures as well as words to label the rows or columns.

USE TECHNOLOGY
Computer

A computer has many uses. The Internet connects your computer to many other computers around the world, so you can collect all kinds of information. You can use a computer to show this information and write reports. Best of all you can use a computer to explore, discover, and learn.

You can also get information from CD-ROMs. They are computer disks that can hold large amounts of information. You can fit a whole encyclopedia on one CD-ROM.

Use Computers for a Project
Here is how one group of students uses computers as they work on a weather project.

1. The students use instruments to measure temperature, wind speed, wind direction, and other parts of the weather. They input this information, or data, into the computer. The students keep the data in a table. This helps them compare the data from one day to the next.

2. The teacher finds out that another group of students in a town 200 kilometers to the west is also doing a weather project. The two groups use the Internet to talk to each other and share data. When a storm happens in the town to the west, that group tells the other group that it's coming its way.

4. Meanwhile some students go to the library to gather more information from a CD-ROM disk. The CD-ROM has an encyclopedia that includes movie clips with sound. The clips give examples of different kinds of storms.

5. The students have kept all their information in a folder called Weather Project. Now they use that information to write a report about the weather. On the computer they can move paragraphs, add words, take out words, put in diagrams, and draw their own weather maps. Then they print the report in color.

3. The students want to find out more. They decide to stay on the Internet and send questions to a local TV weather forecaster. She has a Web site and answers questions from students every day.

USE TECHNOLOGY

Calculator

Sometimes after you make measurements, you have to add or subtract your numbers. A calculator helps you do this.

Add and Subtract Rainfall Amounts

The table shows the amount of rain that fell in a town each week during the summer. The amounts are given in centimeters (cm). Use a calculator to find the total amount of rain that fell during the summer.

Week	Rain (cm)
1	3
2	5
3	2
4	0
5	1
6	6
7	4
8	0
9	2
10	2
11	6
12	5

1. Make sure the calculator is on. Press the ON key.
2. To add the numbers, enter a number and press +. Repeat until you enter the last number. Then press =. You do not have to enter the zeroes. Your total should be 36.
3. Suppose you found out that you made a mistake in your measurements. Week 1 should be 2 cm less, Week 6 should be 3 cm less, Week 11 should be 1 cm less, and Week 12 should be 2 cm less. Subtract these numbers from your total. You should have 36 displayed on the calculator. Press – and enter the first number you want to subtract. Repeat until you enter the last number. Then press =. Compare your new total to your classmates' new totals.

HANDBOOK

R26

GLOSSARY

This Glossary will help you to pronounce and understand the meanings of the Science Words introduced in this book. The page number at the end of the definition tells where the word appears.

A

adaptation (ad′əp tā′shən) A characteristic that helps an organism survive in its environment. (p. 364)

antibody (an′ti bod′ē) A chemical made by the immune system to fight a particular disease. (p. 401)

asteroid (as′tə roid′) A small chunk of rock or metal that orbits the Sun. (p. 250)

atmosphere (at′məs fîr′) A layer of gases surrounding a planet. (p. 238)

axis (ak′sis) A real or imaginary line through the center of a spinning object. (p. 197)

B

bacteria (bak tîr′ē ə) One-celled living things. (p. 399)

PRONUNCIATION KEY

a	at	e	end	o	hot	u	up	hw	white	ə	about
ā	ape	ē	me	ō	old	ū	use	ng	song		taken
ä	far	i	it	ô	fork	ü	rule	th	thin		pencil
âr	care	ī	ice	oi	oil	u̇	pull	<u>th</u>	this		lemon
		îr	pierce	ou	out	ûr	turn	zh	measure		circus

′ = primary accent; shows which syllable takes the main stress, such as **kil** in **kilogram** (kil′ə gram′)
′ = secondary accent; shows which syllables take lighter stresses, such as **gram** in **kilogram**

GLOSSARY

C

camouflage (kam′ə fläzh′) An adaptation that allows organisms to blend into their surroundings. (p. 366)

carbohydrate (kär′bō hī′drāt) A substance used by the body as its main source of energy. (p. 412)

carbon dioxide and oxygen cycle (kär′bən dī ok′sīd and ok′sə jən sī′kəl) The exchange of gases between producers and consumers. (p. 344)

cell (sel) 1. Tiny box-like part that is the basic building block of living things. (p. 56) 2. A source of electricity. (p. 184)

cell membrane (sel mem′brān) A thin outer covering of plant and animal cells. (p. 57)

circuit (sûr′kit) The path electricity flows through. (p. 184)

comet (kom′it) A body of ice and rock that orbits the Sun. (p. 250)

communicate (kə mū′ni kāt′) To share information by sending, receiving, and responding to signals. (p. 8)

community (kə mū′ni tē) All the living things in an ecosystem. (p. 324)

competition (kom′pi tish′ən) When one organism works against another to get what it needs to live. (p. 356)

compound (kom′pound) Two or more elements put together. (p. 158)

compound machine (kom′pound mə shēn′) Two or more simple machines put together. (p. 122)

conifer (kon′ə fər) A tree that produces seeds inside of cones. (p. 35)

consumer (kən sü′mər) An organism that eats producers or other consumers. (p. 334)

corona (kə rō′nə) The outermost layer of gases surrounding the Sun. (p. 229)

crater (krā′tər) A hollow area in the ground. (p. 208)

cytoplasm (sī′tə plaz′əm) A clear, jelly-like material that fills plant and animal cells. (p. 57)

D

data (dā′tə) Information. (p. 82)

decomposer (dē′kəm pō′zər) An organism that breaks down dead plant and animal material. (p. 336)

degree (di grē′) The unit of measurement for temperature. (p. 166)

dermis (dûr′mis) The layer of skin just below the epidermis. (p. 390)

development (di vel′əp mənt) The way a living thing changes during its life. (p. 4)

digestion (di jes′chən) The process of breaking down food. (p. 422)

E

earthquake (ûrth′kwāk′) A sudden movement in the rocks that make up Earth's crust. (p. 282)

eclipse (i klips′) When one object passes into the shadow of another object. (p. 218)

ecosystem (ek′ō sis′təm) All the living and nonliving things in an environment. (p. 324)

element (el′ə mənt) A building block of matter. (p. 157)

PRONUNCIATION KEY

a **at**; ā **ape**; ä **far**; âr **care**; e **end**; ē **me**; i **it**; ī **ice**; îr **pierce**; o **hot**; ō **old**; ô **fork**; oi **oil**; ou **out**; u **up**; ū **use**; ü **rule**; ù **pull**; ûr **turn**; hw **white**; ng **song**; th **thin**; <u>th</u> **th**is; zh **measure**; ə **a**bout, tak**e**n, penc**i**l, lem**o**n, circ**u**s

embryo (em′brē ō) A young organism that is just beginning to grow. (p. 34)

endangered (en dān′jərd) In danger of becoming extinct. (p. 378)

energy (en′ər jē) The ability to do work. (p. 14, 101)

energy pyramid (en′ər jē pir′ə mid′) A diagram that shows how energy is used in an ecosystem. (p. 339)

environment (en vī′rən mənt) The things that make up an area, such as the land, water, and air. (p. 6)

epidermis (ep′ə dûr′mis) The outer layer of skin. (p. 388)

erosion (i rō′zhən) The process that occurs when weathered materials are carried away. (p. 272)

extinct (ek stingkt′) When there are no more of a certain plant or animal. (p. 378)

F

fats (fatz) Substances used by the body as long-lasting sources of energy. (p. 413)

fertilizer (fûr′tə lī′zər) A substance used to keep plants healthy. (p. 311)

fiber (fī′bər) Material that helps move wastes through the body. (p. 414)

flowering plant (flou′ər ing plant) A plant that produces seeds inside of flowers. (p. 35)

food chain (füd chān) A series of organisms that depend on one another for food. (p. 334)

food web (füd web) Several food chains that are connected. (p. 338)

force (fôrs) A push or pull. (p. 78)

friction (frik′shən) A force that occurs when one object rubs against another. (p. 90)

fuel (fū′əl) Something burned to provide heat or power. (p. 230)

G

gas (gas) Matter that has no definite shape or volume. (p. 142)

germinate (jûr′mə nāt) To begin growing. (p. 34)

glacier (glā′shər) A large mass of ice in motion. (p. 272)

gland (gland) A part of the body that makes substances the body needs. (p. 389)

gravity (grav′i tē) The pulling force between two objects. (p. 80)

H

habitat (hab′i tat′) The place where a plant or animal naturally lives and grows. (p. 324)

heat (hēt) A form of energy that makes things warmer. (p. 166)

helper T-cells (hel′pər tē selz) White blood cells that send signals to warn that germs have invaded the body. (p. 401)

hibernate (hī′bər nāt′) To rest or sleep through the cold winter. (p. 18)

host (hōst) The organism a parasite lives in or on. (p. 347)

hurricane (hûr′i kān′) A violent storm with strong winds and heavy rains. (p. 280)

PRONUNCIATION KEY

a at; ā ape; ä far; âr care; e end; ē me; i it; ī ice; îr pierce; o hot; ō old; ô fork; oi oil; ou out; u up; ū use; ü rule; ú pull; ûr turn; hw white; ng song; th thin; <u>th</u> this; zh measure; ə about, taken, pencil, lemon, circus

I

immune system (i mūn′ sis′təm) All the body parts and activities that fight diseases. (p. 403)

immunity (i mū′ni tē) The body's ability to fight diseases caused by germs. (p. 403)

inclined plane (in klīnd′ plān) A flat surface that is raised at one end. (p. 118)

inherited trait (in her′i təd trāt) A characteristic that comes from your parents. (p. 28)

insulator (in′sə lā′tər) A material that heat doesn't travel through easily. (p. 170)

L

landform (land′fôrm′) A feature on the surface of Earth. (p. 264)

large intestine (lärj in tes′tin) Part of the body that removes water from undigested food. (p. 425)

learned trait (lûrnd trāt) Something that you are taught or learn from experience. (p. 28)

lens (lenz) A curved piece of glass. (p. 240)

lever (lev′ər) A straight bar that moves on a fixed point. (p. 109)

life cycle (līf sī′kəl) All the stages in an organism's life. (p. 24)

liquid (lik′wid) Matter that has a definite volume, but not a definite shape. (p. 142)

lunar eclipse (lü′nər i klips′) When Earth's shadow blocks the Moon. (p. 219)

M

machine (mə shēn′) A tool that makes work easier to do. (p. 108)

magnetism (mag′ni tiz′əm) The property of an object that makes it attract iron. (p. 154)

mass (mas) How much matter is in an object. (p. 133)

matter (mat′ər) What makes up an object. (p. 80)

melanin (mel′ə nin) A substance that gives skin its color. (p. 388)

metal (met′əl) A shiny material found in the ground. (p. 154)

metamorphosis (met′ə môr′fə sis) A change in the body form of an organism. (p. 25)

migrate (mī′grāt) To move to another place. (p. 18)

mineral (min′ə rəl) A substance found in nature that is not a plant or an animal. (pp. 49, 260)

mixture (miks′chər) Different types of matter mixed together. (p. 147)

motion (mō′shən) A change of position. (p. 70)

N

natural resource (nach′ər əl rē′sôrs′) A material on Earth that is necessary or useful to people. (p. 292)

nerve cells (nûrv selz) Cells that carry messages to and from all parts of the body. (p. 390)

newton (nü′tən) The unit used to measure pushes and pulls. (p. 78)

niche (nich) The job or role an organism has in an ecosystem. (p. 358)

PRONUNCIATION KEY

a at; ā ape; ä far; âr care; e end; ē me; i it; ī ice; îr pierce; o hot; ō old; ô fork; oi oil; ou out; u up; ū use; ü rule; u̇ pull; ûr turn; hw white; ng song; th thin; th this; zh measure; ə about, taken, pencil, lemon, circus

nonrenewable resource (non′ri nü′ə bəl rē′sôrs′) A resource that cannot be reused or replaced in a useful amount of time. (p. 302)

nucleus (nü′klē əs) A main control center found in plant and animal cells. (p. 57)

nutrient (nüt′rē ənt) A substance that your body needs for energy and growth. (p. 412)

O

opaque (ō pāk′) Does not allow light to pass through. (p. 176)

orbit (ôr′bit) The path an object follows as it revolves. (p. 198)

organ (ôr′gən) A group of tissues that work together. (p. 58)

organism (ôr′gə niz′əm) A living thing. (p. 4)

oxygen (ok′sə jən) A gas that is in air and water. (p. 16)

P

parasite (par′ə sīt) An organism that lives in or on another organism. (p. 347)

perish (per′ish) To not survive. (p. 377)

phase (fāz) Apparent change in the Moon's shape. (p. 207)

plain (plān) A large area of land with few hills. (p. 264)

planet (plan′it) A satellite of the Sun. (p. 228)

plateau (pla tō′) A flat area of land that rises above the land that surrounds it. (p. 265)

pollution (pə lü′shən) What happens when harmful substances get into water, air, or land. (p. 310)

population (pop′yə lā′shən) All the members of a certain type of living thing in an area. (p. 324)

pore (pôr) A tiny opening in the skin. (p. 391)

position (pə zish′ən) The location of an object. (p. 68)

pound (pound) The unit used to measure force and weight in the English system of measurement. (p. 81)

predator (pred′ə tər) An animal that hunts other animals for food. (p. 356)

prey (prā) The animal a predator hunts. (p. 356)

producer (prə dü′sər) An organism that makes its own food. (p. 334)

property (prop′ər tē) A characteristic of something. (p. 135)

protein (prō′tēn) A substance that the body uses for growth and the repair of cells. (p. 413)

pulley (pul′ē) A simple machine that uses a wheel and a rope. (p. 112)

R

recycle (rē sī′kəl) To treat something so it can be used again. (p. 314)

reduce (ri düs′) To make less of something. (p. 312)

reflect (ri flekt′) To bounce off a surface. (p. 177)

relocate (ri lō′kāt) To find a new home. (p. 377)

PRONUNCIATION KEY

a at; ā ape; ä far; âr care; e end; ē me; i it; ī ice; îr pierce; o hot; ō old; ô fork; oi oil; ou out; u up; ū use; ü rule; u̇ pull; ûr turn; hw white; ng song; th thin; <u>th</u> this; zh measure; ə about, taken, pencil, lemon, circus

renewable resource (ri nü′ə bəl rē′sôrs′) A resource that can be replaced or used over and over again. (p. 296)

reproduction (rē′prə duk′shən) The way organisms make new living things just like themselves. (p. 5)

respond (ri spond′) The way a living thing reacts to changes in its environment. (p. 6)

reuse (v., rē ūz′) To use something again. (p. 314)

revolve (ri volv′) To move in a circle around an object. (p. 198)

rotate (rō′tāt) To turn around. (p. 196)

S

saliva (sə lī′və) A liquid in your mouth that helps soften and break down food. (p. 423)

satellite (sat′ə līt′) An object that orbits another, larger object in space. (p. 206)

screw (skrü) An inclined plane wrapped into a spiral. (p. 120)

simple machine (sim′pəl mə shēn′) A machine with few or no moving parts. (p. 109)

small intestine (smôl in tes′tin) A tube-like part of your body where most digestion takes place. (p. 425)

solar eclipse (sō′lər i klips′) When the Moon's shadow blocks the Sun. (p. 218)

solar system (sō′lər sis′təm) The Sun and all the objects that orbit the Sun. (p. 236)

solid (sol′id) Matter that has a definite shape and volume. (p. 142)

solution (sə lü′shən) A type of mixture that has one or more types of matter spread evenly through another. (p. 148)

speed (spēd) How fast an object moves. (p. 71)

star (stär) A hot sphere of gases that gives off energy. (p. 228)

stomach (stum′ək) Part of your body that has walls made of strong muscles that squeeze and mash food. (p. 424)

sunspot (sun′spot′) A dark area on the Sun's surface. (p. 229)

switch (swich) Opens or closes an electric circuit. (p. 185)

system (sis′təm) A group of parts that work together. (p. 46)

T

taste buds (tāst budz) Thousands of cells on your tongue that send the signals for sweet, sour, bitter, and salty to your brain. (p. 423)

telescope (tel′ə skōp′) A tool that gathers light to make faraway objects appear closer. (p. 240)

temperature (tem′pər ə chər) A measure of how hot or cold something is. (p. 166)

tissue (tish′ü) A group of cells that are alike. (p. 58)

V

vaccine (vak′sēn) A medicine that causes the body to form antibodies against a certain disease. (p. 404)

valley (val′ē) An area of land lying between hills. (p. 264)

PRONUNCIATION KEY

a **at**; ā **ape**; ä **far**; âr **care**; e **end**; ē **me**; i **it**; ī **ice**; îr **pierce**; o **hot**; ō **old**; ô **fork**; oi **oil**; ou **out**; u **up**; ū **use**; ü **rule**; u̇ **pull**; ûr **turn**; hw **white**; ng **song**; th **thin**; <u>th</u> **this**; zh **measure**; ə **about, taken, pencil, lemon, circus**

virus (vī′rəs) A tiny particle that can reproduce only inside a living cell. (p. 399)

vitamin (vīt′ə mən) A substance used by the body for growth. (p. 414)

volcano (vol kā′nō) An opening in the surface of Earth. Melted rock, gases, rock pieces, and dust are forced out of this opening. (p. 283)

volume (vol′ūm) How much space matter takes up. (p. 132)

W

weathering (we<u>th</u>′ər ing) The process that causes rocks to crumble, crack, and break. (p. 270)

wedge (wej) Two inclined planes placed back to back. (p. 119)

weight (wāt) The pull of gravity on an object. (p. 81)

wheel and axle (hwēl and ak′səl) A wheel that turns on a post. (p. 111)

white blood cells (hwīt blud selz) Cells in the blood that fight bacteria and viruses. (p. 400)

work (würk) When a force changes the motion of an object. (p. 100)

INDEX

A

Adaptation, 364–366, 368, 381
AIDS, 406–407
Air, 16, 210, 305
Aluminum, 156–157
Animals, 22-23*, 24-27, 28*–29, 46-50*, 51, 57, 323*, 330-331, 334, 342–343*, 344-348*, 349, 356–359, 362–363*, 364, 374–378*
Antibodies, 401–402*, 404, 408
Area, R2–R3, R11
Asteroids, 250, 253
Atmosphere, 238–239, 253
Ax, 119
Axis, 197–198, 201, 224, 236, 256
Axles, 111, 113, 125

B

Bacteria, 336, 399, 408, 426
Balance, R14–R15, R20–21
Bar graph, 82*, 304*
Basalt, 262
Bear, 26
Biceps, 84–85
Bird, 27
Birth, 24–27
Blood, 391
Body parts, 44–45*, 46–50*, 51, 54–55*, 56*-59
Body system, 58
Body temperature, 391
Bulbs, 38

C

Calculator, R26
Calendar, 212–213
Camera, R10
Camouflage, 366–367*, 381
Carbohydrates, 412, 429
Carbon dioxide, 210, 239, 344–345
Carbon dioxide and oxygen cycle, 344–345, 352
Carver, George Washington, 298–299
Cell, electrical, 183*–184, 189, 192
Cell membrane, 57, 61
Cells, 56*, 57–59, 61
Cellulose, 60
Cell wall, 57
Celsius scale, R2–R3, R17
Centimeter, R2–R3
Chalk, changing, 271*
Changes
    in ecosystems, 372–373*, 374–378*, 379
    in living things, 3*–4, 22–23*, 24–27, 28*–29
    in matter, 144–145, 150–151, 168–169*, 177
    in motion, 87*–89, 100
    in rocks, 269*–270
Charts, reading, 27, 71, 143, 177, 184, 313, 357, 364, 368, 416, R23
Chemicals, 270–271*
Chlorine, 158
Chloroplasts, 57
Circuits, 184–185*, 189

Classifying, 50*, 61, 63, 141*–142, 146
Cleaning water, 311*
Clock, R12
Closed circuit, 184
Coal, 230, 302–303
Comets, 250, 253
Communicating, 8, 42, 146*, 191
Community, 324, 328*, 350–352
Comparing the Sun and Moon, 231*
Compass, R8, R10
Competition, 354–355*, 356–358*, 359, 361, 381
Compost, 39
Compound machines, 122–123, 125
Compounds, 158–159, 162
Computer, R24–R25
Conglomerate, 262
Conifers, 35–36, 42
Conservation, 188, 308–309*, 310–311*, 312–315, 360-361
Consumers, 334, 337*, 344–345, 352
Contour farming, 266–267
Contraction, 169
Controlling electrical flow, 185
Controlling experiments, 178*
Copper, 156–157
Corona, 229, 231, 253
Craters, 208, 210, 224, 239
Cuttings, 38
Cytoplasm, 57, 61

* Indicates an activity related to this topic.

INDEX

Day – Food

D

Day, 194–195*, 196
Death, 24
Decomposers, 336–337*, 352
Defining terms, 328*
Degrees, 166, 189, R2–R3
Dermis, 390–391, 408
Designing experiments, 45*, 63, 107*, 127, 141*, 191, 245*, 255, 279*, 319, 363*, 383, 387*, 431
Development, 4, 9, 24, 26–27, 36
Diagrams, reading, 4, 16, 35, 57, 70, 72, 79, 88, 102, 109, 136, 167, 189, 199, 206, 219, 230, 237, 263, 271, 283, 303, 325, 327, 335, 376, 389, 401, 423
Digestion, 409, 422*–427, 429
Digestive system, 426–427, 432
Disease, 381, 388–389, 396–397*, 398–402*, 403–405
Distance, 67*, 71, 217*, R2–R3
Dust bowl, 284

E

Earth, 104–105, 194–195*, 196–200*, 201, 206–211, 214–215*, 216–222, 227*–228, 234–235*, 236–241, 258–259*, 260–261*, 262–265, 268–269*, 270–271*, 272–273*, 274–275, 278–279*, 280, 282–285, 290–291*, 292–295*, 296–297
Earthquakes, 104–105, 282–283
Earth's surface, 264–265, 268–269*, 270–271*, 272–273*, 274–275, 278–279*, 280–281*, 282–285, 290–291*, 292–294, 296–297
 changes to, 268–269*, 278–279*, 280–281*, 282–283
Eclipse, 218–224
Ecosystems, 323*, 324–327, 328*–331, 336–337*, 338–339, 343*–344, 357, 358*–359, 368, 372–373*, 374–377, 378*–379
Eggs, 30–31
Electrical flow, 185*
Electricity, 182–183*, 184–185*, 186–189, 192
Elements, 157–159, 160–162
Embryo, 34, 42
Endangered species, 378*–379, 381
Energy, 14, 16, 42, 49, 101*–102, 104–105, 143, 145, 166, 168, 171, 172–173, 176–177, 180–181, 226–227*, 229–230, 233, 302, 304*, 336, 360–361, 412
Energy pyramid, 339
Energy survey, 304*
Energy transformation, 336
English system of measures, R2–R3
Environment, 6–7*, 8–9, 18, 19, 48, 274, 322–323*, 324–328*, 362–363*, 364–367*, 368–369
Epidermis, 389, 390, 397*, 408
Epiglottis, 428
Equal forces, 89
Erosion, 266–267, 272, 273*, 274–279*, 280–281*, 296
Esophagus, 424
Evaporation, 145
Expansion, 169
Experimenting, 7*
Experiments, 7*, 45*, 63, 64, 107*, 127, 141*, 178*, 191, 245*, 255, 279*, 319, 383, 387*, 431. *See also* Explore activities *and* Quick Labs.
Explore activities, 3*, 13*, 23*, 33*, 45*, 55*, 67*, 77*, 87*, 99*, 107*, 117*, 131*, 141*, 153*, 165*, 175*, 183*, 195*, 205*, 215*, 227*, 235*, 245*, 259*, 279*, 291*, 301*, 309*, 311*, 323*, 333*, 343*, 355*, 373*, 387*, 397*, 411*, 421*
Extinction, 378*–379, 381

F

Fahrenheit scale, R2–R3, R17
Fats, 411*– 413, 429
Fertilizers, 311
Fiber, 414, 429
Flashlight, 185*
Flowering plants, 35, 42
Food, 332–333*, 340–341, 344–345, 410–411*, 412–415*, 416–417, 418–419, 420–421*, 422*–428

R40   * Indicates an activity related to this topic.

Food chains, 334–335, 338, 352
Food pyramid, 416
Food webs, 338, 352
Forces, 77*–81, 86–87*, 88–92*, 93, 94, 100, 109–110, 112, 119, R3, R16
Forest community, 328*
Forming a hypothesis, 273*
Formulating a model, 402*
Fortified cereal, 155*
Freeze-dried foods, 418–419
Friction, 90–91, 92*, 93, 94–95, 102, 104–105, 171
Frog, 25, 30–31
Fuels, 230, 233, 253, 302
Fulcrum, 109–110*
Fungi, 336

G

Gases, 16, 49, 142–143, 145–146*, 147–149, 150–151, 157, 169*, 210, 230–231, 248, 249, 250, 283, 302, 305, 344–345
Germination, 34, 37, 42
Germs, 396–397*, 398–402*, 403–405, 424, 426
Gills, 16
Glaciers, 272
Gland, 389, 390, 408, 424
Global positioning system, 74–75
Gneiss, 263
Gold, 156–157
Gram, 134, R2–R3
Granite, 260, 263
Graphs, R20–R21
Gravity, 80–81, 82*, 83, 136–137, 210, 220

Great Dark Spot, 249
Great Red Spot, 246
Growth and change, 24–27, 28*, 29

H

Habitat, 324–329, 344–345, 374–378*
Halley's comet, 252
Hand lens, R6
Heat, 164–165*, 166–169*, 170–171, 189, 230, 238
Helper T-cells, 400
Hibernation, 18
HIV virus, 406–407
Homes, 330–331
Host, 347, 352
Hurricanes, 280–281*, 286–289
Hydrogen, 158

I

Ideas, using, 61, 63, 125, 127, 189, 191, 253, 255, 317, 319, 381, 383, 429, 431
Identifying properties, 367*
Immune system, 403–404, 407–408
Immunity, 403, 408
Immunodeficiency, 406–407
Inclined plane, 118, 120, 125
Inferring, 247*, 253, 255
Inherited traits, 28*, 42
Inner planets, 234–235*, 236–237
Insects, 45*–46
Insulators, 170, 189
Interpreting data, 82*, 127*, 415*, 429
Investigating

ecosystems, 373*
food, 333*, 411*, 421*
forces, 77*
heat, 165*
light, 175*, 183*
living things, 3*, 13*, 23*, 55*, 343*
mining, 301*
motion, 87*
night and day, 195*
planets, 235*
plants and animals, 323*, 355*
protection against disease, 397*
rocks, 259*, 269*
shape of the Moon, 205*
size of the Sun and Moon, 215*
soil, 291*
speed, 67*
Sun's energy, 227*
what organisms need, 13*
work, 99*, 117*
Iron, 155*, 158, 187*

J

Journal writing, 63, 127, 191, 255, 319, 383, 431
Jupiter, 83, 236–237, 244–245*, 246–247*

K

Kidneys, 16
Kilogram (kg), 134

L

Landforms, 264, 268, 283
Landslides, 282–283
Large intestine, 425–426, 429

INDEX

Learned traits – Motion

Learned traits, 24, 42
Length, R2–R3, R11
Lenses, 240*
Levers, 109–110*, 113, 114–115, 125, 128
Life cycles, 24–27, 28*–29, 32–33*, 34–37*
Light, 37*, 174–175*, 176–177, 178*–179, 183*–184, 192, 230, 247
Light waves, 180–181
Limestone, 260, 262–263
Liquid, 16, 49, 140–141*, 142–143, 145–146*, 147, 148–149, 150–151, 157, 162, 168, 178*
Lists, identifying, 61, 125
Living things, 2–3*, 4–7*, 8–11, 13*–15*, 16–19, 22–23*, 24–28*, 29, 33*–37*, 38–45*, 46–50*, 51–55*, 56*, 57–61, 322–323*, 324–328*, 329–337, 342–343*, 344–348*, 349–352, 354–355*, 356–358*, 359–363*, 364–369
  changes in, 3*–4, 22–23*, 24–27, 28*–29
  competition of, 354–355*, 356–358*, 359–361
  environment of, 6–7*, 8–9, 18–19, 48, 274, 322–323*, 324–328*, 362–363*, 364–367*, 368–369
  features of, 4–5, 6–7*, 8–9
  life cycle of, 24–27, 28*–29, 32–33*, 34–37*
  needs of, 13*–15*, 16–19, 342–343*, 344–348*, 349
  parts of, 44–45*, 46–49, 50*–51, 54–55*, 56*, 57–59
  places to live, 322–323*, 324–328*, 329–337*
  recycling, 336–337
  roles for, 342–343*, 344–348*, 349–352
  survival of, 362–363*, 364–368*, 369

L

Length, R2–R3, R11
Lizards, 370–371
Load, 109–110, 112
Lunar calendar, 212–213
Lunar eclipse, 219, 222–224
Lungs, 16

M

Machines, 108–110*, 111–113, 116–117*, 118–121*, 122–125
Magnetism, 154, 162
Magnets, 152–153*, 154
Maps, 72, R10, R22
Marble, 263
Mars, 83, 236–237, 239, 242–243
Mass, 133, 134*, 135–136, 138–139, 210
  measuring, 134*, 136, 138–139, 295, R2–R3, R14–R15
Matter, 80–82*, 83, 132–133, 134*–137, 141*–146*, 147–153*, 168–169, 177
  building blocks of, 152–153*, 154–155*
  changes in, 144–145, 150–151, 168–169*, 177
  classifying, 141*–146*
  forms of, 140–141*, 142–143, 147–149
  gravity and, 80, 81–82*, 83
  heat and, 164–165*, 166–169*, 170–171, 230, 238
  properties of, 134*–135, 136–137, 140
Measurement, 134*, 136, 138–139, 295*, R2–R3
Melanin, 388–389, 408
Mercury, 236–237, 238
Metals, 154–155*, 156, 162
Metamorphosis, 25–26, 42
Meters, R2
Metric system, R2–R3
Microscope, 56*, R7
Migration, 18, 42
Millimeter, R2–R3
Minerals, 49, 61, 260–261*, 293, 414
Mineral scratch test, 261
Mining, 301*–302
Mixtures, 147–149, 158, 162
Models, making, 402*, 408
Monarch butterfly, 2–21, 24
Moon, 204–205*, 206–208, 209*, 210, 212–213, 214, 215*–216, 217*–222, 231*, 236–23, R9
Motion, 70–73, 86–87*, 88–92*, 93, 94, 99*, 100–101*, 102–103
  changes in, 87*–89, 100
  defining, 70–71
  forces in, 77*–81, 86–87*, 88–92*, 93, 94, 100, 109–110, 112, 119
  friction, 90–91, 92*, 93, 94–95, 102, 104–105, 171

R42   * Indicates an activity related to this topic.

of planets, 198, 206–207, 218, 224, 235–238, 239, 246, 248, 249, 250
Movement, 66–67*, 68–73, 86–87*, 88–92*, 93, 99*, 100–101*, 102–103, 235*–236
Moving parts, 46, 48
Moving things, 66–67*, 68–73
   position, 68–69*, 70–72
   speed, 67*, 71
Muscles, 84–85
Music machines, 114–115

N

Natural gas, 302, 304
Natural resources, 292–297, 300–301*, 302–304*, 305, 360–361
Nectar, 35
Neptune, 236–237, 249
Nerve cells, 390–391, 408
Nerves, 391
Newton, 78, 136, R3
Niche, 358*–359, 381
Night, 194–195*, 196
Nonrenewable resources, 302–303, 360–361
North pole, 197–198
Nucleus, 57, 61
Numbers, using, 121*
Nutrients, 34, 39, 412–415*, 416–417
Nutrition, 409, 415*
Nutrition label, 415

O

Objects, 130–131*, 132–134*, 136–137
Observations, making, 247*, 328*, 367*
Obsidian, 262
Oil, 230, 304
Opaque, 176, 189
Open circuit, 184
Orbit, 198, 206–207, 218, 224, 235–238, 239, 246, 248, 249, 250
Organ, 58, 61
Organisms, 4–5, 6–7*, 8–9, 12–13*, 14–19, 22–23*, 24–28*, 29, 44–45*, 46–50*, 51, 54–55*, 56*–59, 323*, 324–327, 328*–331, 336–337*, 338–339, 343*–344, 345–348*, 349, 357, 358*–359, 362–363*, 364–367*, 368–369, 372–373*, 374–377, 378*–379
   body parts of, 44–45*, 46–50*, 51, 54–55*, 56*–59
   changes in, 3*–4, 22–23*, 24–27, 28*–29
   decomposers, 336–337*
   ecosystems of, 323*, 324–327, 328*–331, 336–337*, 338–339, 343*–344, 357, 358*–359, 368, 372–373*, 374–377, 378*–379
   environment, 322–323*
   features of, 3*–7*, 8–9
   needs of, 13*–15*, 16–19, 342–343*, 344–348*, 349
   organization of, 58
   response of, 6–7*, 8–9, 17–18, 48
   survival characteristics, 362–363*, 364–367*, 368–369, 377–378*
Outer covering, 47
Oxygen, 16, 158, 210, 344–345

P

Parasite, 347, 352
Partial eclipse, 220
Parts of living things, 44–45*, 46–49, 50*–51, 54–55*, 56*–59
   parts of parts, 46
   parts that get information, 46, 48
   parts that move, 46, 48
   parts that protect and support, 46–47, 50*
   parts that take in materials, 46, 49
   smaller, 54–55*, 56*, 57–59
Patterns, using, 209*
Periodic table, 160–161
Perishing, 377–378
Phases, of the Moon, 206–207, 209*, 211, 224
Piano, 114–115
Plain, 264–265
Planetary rings, 246–249
Planets, 228, 234–235*, 236–239, 240*, 241–246, 247*, 248–251, 253
Plant life cycle, 33*–34, 36
Plants, 32–33*, 34–37*, 38–39, 40–41, 46, 52–53, 57, 60, 64, 323*, 330–331, 334, 342–343*, 344–348*, 349, 355*–359, 364–365, 374–375
Plastics, 306–307
Plateau, 265

INDEX

Plow – Small intestine

Plow, 119
Pluto, 236–237, 249
Pollen, 35–36
Pollination, 35–36
Pollution, 308–309*, 310–311*, 312, 316
Pond ecosystem, 326
Population, 324, 344–347, 352
Pores, 391, 408
Position, 68–69*, 70–72
Pounds, 81, R2–R3
Predator, 356, 360–361, 373*, 381
Predicting, 63, 191, 209*, 255
Prey, 356, 373*, 381
Problems and puzzles, 42, 61, 64, 96, 125, 128, 162, 224, 253, 256, 288, 317, 320
Problem solving, 64, 128, 192, 256, 320, 384, 432
Producers, 334, 337*, 344–345, 352
Prominences, 231*
Properties, 135–137, 140–141*, 142–143, 147, 150–151, 153*, 154–155*, 158, 162, 175*, 294, 367*
Proteins, 413
Pulley, 112–113, 125, 128
Pulls, 76–77*, 78–81, 82*, 83, 109
Pushes, 76–77*, 78–81, 82*, 83, 109

Q

Quick Lab, 15*, 28*, 37*, 56*, 69*, 92*, 101*, 110*, 134*, 155*, 169*, 185*, 200*, 217*, 231*, 240*, 261*, 271*, 281*, 337*, 348*, 358*, 378*, 392*, 422*

R

Rabbits, 381
Radar, 288–289
Radiating, 220
Rain forest, 357
Ramp, 117*–118
Rate, R3
Recycling, 314–315, 336–337, R5
Reducing, 312
Reflection, 176–177, 189, 208
Refrigeration, 172–173
Relocation, 377–378
Renewable resources, 296, 300, 305
Reproduction, 5, 9, 24, 33*, 34–36, 38, 42, 348*, 365
Resource conservation, 308–309*, 310–315
Resources, 292–297, 300–301*, 302–304*, 305, 308–309*, 310–315, 360–361
Response of organisms, 6–7*, 8–9, 17–18, 48
Reuse, 314–315
Revolution, 198–199, 210, 224, 236, 239, 248
Rock climbing, 94–95
Rocks, 94–95, 259*–260, 261*–263, 269*, 270, 272–273*, 274–275, 290–291*
    changes in, 269*–270
    comparing, 259*–261*
    forming, 262–263
Rotation, 196, 210, 224, 236

Rust, 158

S

Safety rules, R4–R5
Saliva, 423
Sandstone, 262
Satellite, 206, 220, 224, 239
Saturn, 236–237, 248, 256
Schist, 263
Science Journal, 63, 127, 191, 255, 319, 383, 431. *See also* Explore Activities, Quick Lab, Skill Builders.
Screw, 120, 121*, 125
Seasons, 198
Seeds, 35–37*, 348*, 384
Seismic waves, 180–181
Serving size, 415
Shade, 37*
Shadow, 218–220
Shale, 262, 263
SI (International System) measures, R2–R3
Sign language, 14–15
Silver, 156
Simple machines, 108–109, 110*–113, 116–117*, 119–121*, 122–125
Skill builders, 7*, 50*, 82*, 121*, 146*, 178*, 209*, 247*, 273*, 295*, 328*, 367*, 402*, 415*
Skills, using, 7*, 50*, 61, 63, 82*, 125, 127, 146*, 178*, 189, 191, 209*, 247*, 253, 255, 273*, 295*, 317, 319, 328*, 367*, 381, 383, 429, 431
Skin, 386–387*, 388–391, 392*–395, 397*
Slate, 263
Small intestine, 425, 429

R44   * Indicates an activity related to this topic.

Sodium, 158
Sodium chloride, 158
Soil, 39, 165*, 266–267, 290–291*, 292–295*, 296, 384
   layers of, 293
   properties of, 294
Soil formation, 293
Solar cells, 233
Solar eclipse, 218, 220, 222–224
Solar energy, 360–361
Solar flare, 231
Solar storm, 231
Solar system, 236–240*, 244–245*, 246–247*, 248–251, 253, 256
Solar wind, 250
Solids, 16, 49, 140–141*, 142–143, 145–146*, 147, 150–151, 157, 162
Solution, 148, 162
Sound waves, 180–181
South pole, 197
Space, 131*–132, 137
Space foods, 418–419
Space probe, 238, 242–243, 256
Speed, 67*, 71
Sphere, 205*–206
Spores, 38
Spring scale, 77*–78, R16
Stars, 202–203, 228, 253
Star time, 202–203
Steel, 155*–156
Stomach, 424–425, 429
Stone symbols, 266–267
Stopwatch, R12
Subsoil, 293
Summarizing, 253, 429
Sun, 194–195*, 196–200*, 203, 208, 214, 215*–216, 217*–218, 220–223, 227*–233, 235*, 237*–239, 246, 249, 250, 394–395
Sundial, 199*
Sun's energy, 226–227*, 228, 229–231
Sunspots, 229, 231, 253
Survey, 304*
Survival, 362–363*, 364–367*, 368–369, 377–378*
Switch, electrical, 185, 189
System, R19
   electrical, 184
Systems of living things, 46, 61

T

Tables, making, 50*, 146*, R23
Tape recorder, R10
Taste buds, 423, 429
Technology in science, 74–75, 172–173, 288–289
Telescope, 240*–241, 253, R9
Temperate forest, 357
Temperature, 166, 227*–228, 281*, 391, R2–R3, R17–R18
Terrace farming, 266–267
Thermometer, 166, 168
Tilted axis, 197–198, 201, 248, 256
Time, 67*, 71, 202–203, 212–213
Tissue, 58, 61
Topsoil, 293
Total eclipse, 220
Trash, 313–314
Triceps, 84–85
Tubers, 38

U

Ultraviolet light, 394–395
Unbalanced forces, 89
Unequal forces, 89
Uranus, 236–237, 248
Using numbers, 121*
Using observations, 247
Using patterns, 209*
Using variables, 178*

V

Vaccines, 404, 408
Valley, 264–265
Variables, using, 178*, 189
Venus, 236–237, 238
Viruses, 381, 399, 406–408
Vitamins, 414, 429
Volcanic eruption, 372
Volcanoes, 239, 283
Volume, 131*–132, 133, 135, 245*–246, 295*, R2–R3, R13

W

Wastes, 15*, 16, 49, 391, 425
Water, 15*, 158, 165*, 210, 295*, 305, 308–309*, 310–311*, 360–361, 414
Water filtration model, 311*
Water pollution, 308–309*, 310–311*, 316
Water transport system, 58
Water vapor, 145
Weather, 281
Weathering, 270, 272, 273*–275, 278–279*, 280–281*
Weather satellites, 288–289
Wedge, 119, 125

Weight – Year

Weight, 81, 82*, 83, 136, R2–R3, R16
Wheels, 111–113, 125
White blood cells, 400, 403, 408
Wind energy, 360–361
Windpipe, 428
Work, 98–99*, 100–101*, 106–107*, 108, 110*, 111–113, 116–117*, 118–121*, 122–123, 125

Y

Year, 198

CREDITS

Design & Production: Kirchoff/Wohlberg, Inc.

Maps: Geosystems.

Transvision: Ken Karp (photography); Michael Maydak (illustration).

Illustrations: Ken Batelman p.428; Ka Botzis: pp. 271, 274, 293, 325, 368, 376; Elizabeth Callen: 360; Barbara Cousins: pp. 85, 423, 424, 425; Steve Cowden pp. 350–351; Marie Dauenheimer: pp. 388–389, 390–391, 399, 400–401; Michael DiGiorgio: pp. 328, 335, 339, 364; Jeff Fagan: pp. 12, 58, 88, 89, 91, 101, 102; Lee Glynn: pp. 15, 71, 72, 82, 83, 136, 159, 230, 256, 313, 352, 357, 384, 398, 416, 432; Kristen Goeters: p. 137; Colin Hayes: p. 173 Handbook pp. R7, R11, R13, R20–R23; Nathan Jarvis: pp. 68, 69, 70; Matt Kania: pp. 264, 283; Virge Kask: pp. 14, 26–27; Fiona King: 222, 223. Tom Leonard: pp. 16, 57, 81, 90, 196, 197, 200, 208, 236–237; Olivia: Handbook pp. R2–R4, R9, R10, R13, R16–R19, R21, R23–R25; Sharron O'Neil: pp. 4, 34, 35, 36, 40, 60, 64, 288, 303, 317, 320; Pat Rasch: pp. 79, 80, 118, 119, 120, 121, 128; Rob Schuster: pp. 115, 185, 186, 192, 198–199, 206–207, 216, 218, 219, 244; Casey Shain: p. 304; Wendy Smith: pp. 338, 344, 326–327; Matt Straub: pp. 42, 61, 96, 125, 162, 166, 189, 224, 228, 253, 317, 352, 381, 408, 429; Ted Williams: pp. 154, 156, 167, 178, 182, 184; Jonathan Wright: pp. 110, 111, 113.

Photography Credits:

Contents: iii: Bob & Clara Calhoun/Bruce Coleman, Inc. iv: inset, Bob Winsett/Corbis; FPG. v: Richard Megna/Fundamental Photographs. vi: ESA/Science Photo Library. vii: Roger Werth/Woodfin Camp & Associates, Inc. viii: Gregory Ochocki/Photo Researchers, Inc. ix: Walter Bibikow/FPG.

National Geographic Invitation to Science: S2: Emory Kristof; inset, Harriet Ballard. S3: t. Woods Hole Oceanographic Institution; b. Jonathan Blair.

Be a Scientist: S5: David Mager. S6: NASA. S7: t. Corbis; b, Francois Gohier/Photo Researchers, Inc. S8: l, Jonathan Blair/Woodfin Camp & Associates; r, Wards SCI/Science Source/Photo Researchers, Inc. S11: NASA. S12: John Sanford/Science Photo Library/Photo Researchers, Inc. S13: t, b, NASA. S14: Michael Marten/Science Photo Library/Photo Researchers, Inc. S15: Peter Beck/The Stock Market. S16: l, K. Preuss/The Image Works; r, Richard A. Cooke III/Tony Stone Images. S17: Jean Miele/The Stock Market.

Unit 1: 1: Dieter & Mary Plage/Bruce Coleman, Inc.; Randy Morse/Animals Animals, inset b.r. 2: Richard Nowitz/FPG. 3: Ken Karp. 5: Ken Karp, t.r.; R. Calentine/Visuals Unlimited, b. 6: Barry L. Runk/Grant Heilman, b.l.; Runk/Schoenberger/Grant Heilman, b.r. 7: Ken Karp. 8: Sullivan & Rogers/Bruce Coleman, Inc., t.r.; Tom J. Ulrich/Visuals Unlimited, b.c. 9: Cart Roessler/Animals Animals. 10: Ronald H. Cohn. H. S. Terrence 11: Animals Animals. 13: Ken Karp. 15: Ken Karp. 17: C. Bradley Simmons/Bruce Coleman, Inc., t.r.; Jerry Cooke/Animals Animals, b. 18: Jim Zipp/Photo Researchers, Inc., c.; Kim Taylor/Bruce Coleman, Inc., r.; Lefever/Grushow/Grant Heilman, l. 19: Arthur Tilley/FPG. 20: Ken Lucas/Visuals Unlimited, l. 20–21: Skip Moody/Dembinsky Photo Assoc. 21: The Blake School, t. 22: Tim Davis/Zipp/Photo Researchers, Inc. 23: Ken Karp. 24: Dwight R. Kuhn, t.l.; Glenn M. Oliver/Visuals Unlimited, t.c.; Pat Lynch/Zipp/Photo Researchers, Inc., t.r.; Robert P. Carr/Bruce Coleman, Inc., b.l. 25: John Mielcarek/Dembinsky Photo Assoc., b.l.; Nuridsany et Perennou/Zipp/Photo Researchers, Inc. t.l.; Robert L. Dunne/Bruce Coleman, Inc., t.r.; Sharon Cummings/Dembinsky Photo Assoc., b.r.. 26: Henry Ausloos/Animals Animals. 28: Debra P. Hershkowitz/Bruce Coleman, Inc., t.l.; Ken Karp, b.r. 29: Rhoda Sidney/PhotoEdit. 30: Bill Banaszewski/Visuals Unlimited, inset. 30–31: J.C. Carton/Bruce Coleman, Inc., bkgrd. 32: Toyohiro Yamada/FPG. 33: Ken Karp. 34: Inga Spence/Visuals Unlimited, t.c.; Patti Murray/Animals Animals, l. 36: George F. Mobley, l. 37: Bill Bachman/Photo Researchers, Inc. 38: D. Cavagnaro/Visuals Unlimited, t.l.; Dwight R. Kuhn, b.r..; Dwight R. Kuhn, b.l.; John Lemker/Animals Animals, b.c. 39: Larry Lefever/Grant Heilman. 40–41: Randy Green/FPG, bkgrd.; Stan Osolinski/Dembinsky Photo Assoc. inset t.;. Larry West/FPG, inset b. 41: John M. Roberts/The Stock Market, inset t.; J. H. Robinson/Photo Researchers, Inc., inset b. 43: Superstock; Peter Cade/Tony Stone Images, inset b.r.

44: PhotoDisc., all. 45: Ken Karp. 46: Rob Gage/FPG. 47: Joe McDonald/Animals Animals, b.r.; John Shaw/Bruce Coleman, Inc., t.r. 48: Leonard Rue III/Visuals Unlimited, b.; Robert P. Carr/Bruce Coleman, Inc., t.l. 49: F.C. Millington-TCL/Masterfile, b.r.; Tom McHugh/Photo Researchers, Inc., t.r. 51: Bonnie Kamin/PhotoEdit. 52: Joyce Photographics/Photo Researchers, Inc., t.; Sonya Jacobs/The Stock Market, l. 53: John D. Cunningham/Visuals Unlimited, r.; John Sohlden/Visuals Unlimited, b.l.; Michael T. Stubben/Visuals Unlimited, t.c.; R.J. Erwin/Photo Researchers, Inc., t.l. 54: Ken Karp. 55: Margaret Oechsli/Fundamental Photographs. 56: Dwight R. Kuhn, t.l.; Ken Karp, b.r. 59: Dennis MacDonald/PhotoEdit. 60: Phillip Hayson/Photo Researchers, Inc.

Unit 2: 65: ZEFA Stock Imagery, Inc. 66: Anderson Monkmeyer, b.l.; Dollarhide/Monkmeyer, b.r. 67: Ken Karp, b.r.; Will Hart/PhotoEdit, t.r.. 69: Ken Karp. 70: Barbara Leslie/FPG, b.r.; K.H. Switak/Photo Researchers, Inc., b.l. 71: K. & K. Amman/Bruce Coleman, Inc./PNI. 73: Jacob Taposchaner/FPG. 74: Dan McCoy/Rainbow/PNI. 75: David Young-Wolff/PhotoEdit. 76: Ken Karp. 77: Ken Karp. 78: Ken Karp. 80: RubberBall Productions. 84: Ken Karp. 86: Ken Karp. 87: Ken Karp. 90: NASA. 91: Ken Karp. 92: Ken Karp. 93: Jade Albert/FPG. 94–95: Stephen J. Shaluta, Jr./Dembinsky Photo Assoc. 95: Ken Karp. 97: Chris Salvo/FPG. 98: Camelot/Photonica, b.r.; Jacob Taposchaner/FPG, b.l.; Will & Deni McIntyre/Photo Researchers, Inc., t.c. 99: Ken Karp. 100: Camelot/Photonica, b.l.; Ken Karp, t.l. & m.l. 101: Ken Karp. 103: R. Hutchings/PhotoEdit. 104–105: Ed Degginger/Bruce Coleman, Inc., bkgrd. 105: Jeff Foott/Bruce Coleman, Inc. b. inset; Jonathan Nourok/PhotoEdit, t. inset. 106: Ken Karp. 107: Ken Karp. 109: Ken Karp. 110: Ken Karp, t. & b. 112: Ken Karp. 114–115: The Granger Collection New York. 116: Carl Purcell/Photo Researchers, Inc. 117: Ken Karp. 118: Dollarhide/Monkmeyer. 119: W. Metzen/Bruce Coleman, Inc. 123: Ken Karp. 124: David Mager.

Unit 3: 129: Bkgrd: MMSD Joe Sohm/ChromoSohm. 130: Ken Karp. 131: Ken Karp. inset 132: PhotoDisc. 133: MMSD, m.r.; PhotoDisc, m.c., b.l. & b.r.; Sylvain Grandadam/Photo Researchers, Inc., t.r. 134: Stockbyte. 135: PhotoDisc, b.l.; Ken Karp, t.r. 138: Robert Rathe/NIST; inset, Joe Sohm/Stock, Boston/PNI. 140: Gerry Ellis/ENP Images. 141: Ken Karp. 142: PhotoDisc. 143: PhotoDisc, b.c. & b.r.; Ken Karp. 144: Lawrence Migdale, l. & b.m.; Margerin Studio/FPG, t.r. 145: Ken Karp. 146: Peter Scoones-TCL/Masterfile. 147: PhotoDisc. 148: Ken Karp. 149: Arthur Tilley/FPG. 150: McGraw Hill School Division. 150–151: Ken Karp, insets. 152: Ken Karp. 153: Ken Karp. 154: Leonard Lessin/Peter Arnold, Inc. 155: PhotoDisc, t.r.; Ken Karp, b.r. 156: Telegraph Colour Library/FPG. 157: Ken Karp. 158: PhotoDisc. 160: Stan Osolinski/Dembinsky Photo Assoc. Charles D. Winters/Photo Researchers, Inc.161: t. Mehau Kulyk/Photo Researchers, Inc. b. William Waterfall/The Stock Market. 163: Eric Meola/The Image Bank; Tom Bean, inset b.r. 164: Ken Karp. 165: Ken Karp. 168: Ben Simmons/The Stock Market, l.; Eric Gay/AP/World Wide Photos, b.r. 169: Ken Karp. 170: Nakita Ovsyanikov/Masterfile, b.r.; Robert P. Carr/Bruce Coleman, Inc., l. 171: Ken Karp. 172: Culver Pictures, Inc. 172–173: Gary Buss FPG, bkgrd. 174: Ken Karp. 175: Ken Karp. 176: Ron Thomas. 177: Gary Withey/Bruce Coleman, Inc., b.r.; Jerome Wexler/Photo Researchers, Inc., t.r.; Ken Karp, b.l.; Telegraph Colour Library/FPG, b.c. 179: Tim Davis/Photo Researchers, Inc. 180: Frank Krahmer/Bruce Coleman, Inc., b.l. 181: Telegraph Colour Library/FPG, bkgrd.; Ken Karp, inset. 183: Ken Karp. 188: PhotoDisc bkgrd.; Ken Karp, insets.

Unit 4: 193: NASA/FPG; inset, GSO Images/The Image Bank. 194: George D. Lepp/Photo Researchers, Inc. 195: Ken Karp. 197: Jim Cummins. FPG. 200: Ken Karp. 201: Andy Levin/Photo Researchers, Inc. 202: Michael R. Whelan, inset; Jim Ballard/AllStock/PNI, t. 204–205: Edward R. Degginger/Bruce Coleman, Inc. 205: Ken Karp. 206–207: John Sanford/Science Photo Researchers, Inc. 208–209: NASA. 210: NASA, b.l.; Michael P. Gadomski/Photo Researchers, Inc., t.l. 211: Richard T. Nowitz/Corbis. 212: Chris Dube. 212–213: t. Photo Disc. 213: The Granger Collection New York. 214: Matt Bradley/Bruce Coleman, Inc. 215: Ken Karp. 217: Archive Photos/PNI, b.r.; Ken Karp, b.l. 218: Frank Rossotto/The Stock Market. 219: Rev. Ronald Royer/Photo Researchers, Inc. 220–221: Pekka/Photo Researchers, Inc. 222-223: Visuals Unlimited. 225: Science Photo Library/Photo Researchers, Inc. 226: Mike Yamashita/Woodfin Camp & Associates. 227: Ken Karp. 228: Jerry Schad/Photo Researchers, Inc. 229: Francois Gohier/Photo Researchers, Inc., b.; Jerry Lodriguss/Photo Researchers, Inc., t. 231: Detlev Van/Photo Researchers, Inc., t.; Ken Karp, b. 232: Jim Cummins/FPG. 233: t. NASA/Photo Researchers, Inc. b. Telegraph

R47

Colour Library/FPG 234: Palomar Observatory/Caltech. 235: Ken Karp. 238: NASA/Mark Marten/Photo Researchers, Inc., b.; US Geological/Photo Researchers, Inc., t. 239: NASA/Science Source/Photo Researchers, Inc., b.; US Geological Survey/Photo Researchers, Inc., t. 240: Ken Karp. 241: Mugshots/The Stock Market. 242: A. Ramey Stock Boston, l. NASA/JPL/Corbis; 242–243: USGS/Photo Researchers, Inc., bkgrd. 243: NASA/Corbis, b.r. Photo Researchers, Inc., bkgrd. 245: Ken Karp. 246: Science Photo Library/Photo Researchers, Inc. 247: Ken Karp. 248: NASA, t.; NASA/Mark Marten/Photo Researchers, Inc., b. 249: NASA Science Photo Library/Photo Researchers, Inc., b.; Space Telescope/Photo Researchers, Inc., t. 250–251: Jerry Lodriguss/Photo Researchers, Inc. 251: Nieto/Jerrican/Photo Researchers, Inc. 252: Sam Zarembar/The Image Bank, bkgrd.; The Granger Collection, inset.

Unit 5: 257: ZEFA/Stock Imagery, Inc. 258: PhotoDisc, b.l.; Ann Purcell/Photo Researchers, Inc., b.r.; Jeffrey Myers/FPG., t.r. 259: Ken Karp. 260: Joyce Photographics/Photo Researchers, Inc., b.c.; Ken Karp, t.l., m.l., b.l., b.r. 261: Ken Karp. 262: PhotoDisc, bkgrd; Ken Karp, insets. 263: l. col. from top, Stephen Ogilvy, Ken Karp, Stephen Ogilvy, E.R. Degginger/Photo Researchers, Inc.; r. col. from top, Stephen Ogilvy, Ken Karp, Ken Karp, Charles R. Belinky/Photo Researchers, Inc. 264: Diane Rawson; Photo Researchers, Inc., b.l.; Josef Beck/FPG, m.l.; Tim Davis/Photo Researchers, Inc, b.r. 265: Yann Arthus-Bertrand/Corbis. 266: Art Wolfe/AllStock/PNI, t.; Robert Harding Picture Library, inset; 267: Fergus O'Brien/FPG International, t.; D. E. Cox/Tony Stone Images, m. 268: Francois Gohier/Photo Researchers, Inc. 269: Ken Karp. 270: Keith Kent/Science/Photo Researchers, Inc., bkgrd.; Susan Rayfield/Photo Researchers, Inc., inset l & r. 271: Ken Karp. 272: Farrell Grehan/Photo Researchers, Inc., t.; Ken M. Johns/Photo Researchers, Inc., b. 273: Ken Karp. 274: Dan Guravich/Photo Researchers, Inc. 275: Ralph N. Barrett/Bruce Coleman, Inc. 276: Adam Jones/Photo Researchers, Inc., t.r.; John Sohlden/Visuals Unlimited, b.r.; W. E. Ruth/Bruce Coleman, Inc., b.l. 276–277: PhotoDisc, bkgrd. 277: The National Archives/Corbis, t.r.; Pat Armstrong/Visuals Unlimited, b.l.; Sylvan H. Wittaver/Visuals Unlimited, t.l. 278: Warren Faidley/International Stock. 279: Ken Karp. 280: NASA/GSFC/Photo Researchers, Inc. 281: PhotoDisc. 282: Paul Sakuma/AP/Wide World Photos, b.; Will & Deni McIntyre/Photo Researchers, Inc., t. 283: PhotoDisc. 284: Arthur Rothstein/AP Photo, b.; Sergio Dorantes, t. 285: The Weather Channel. 286: Jeffrey Howe/Visuals Unlimited. 286–287: Telegraph Colour Library/FPG, bkgrd. 287: Frank Rossotto/The Stock Market, t.r.; NOAA/Science Photo Library/Photo Researchers, Inc., m.r.; Dr. Denise M. Stephenson-Hawk, b.r. 289: PhotoDisc, bkgrd.; Stock Imagery, Inc., inset. 290: PhotoDisc, b.r.; Michael P. Gadomski/Photo Researchers, Inc., b.l.; Peter Skinner/Photo Researchers, Inc., t.r. 291: Ken Karp. 292: Craig K. Lorenz/Photo Researchers, Inc. 294: Ken Karp. 295: Ken Karp. 296: Jim Foster/The Stock Market, b.; M. E. Warren/Photo Researchers, Inc., t. 297: Debra P. Hershkowitz/Bruce Coleman, Inc. 298: The National Archives/Corbis, inset. 298–299: John Elk III/Bruce Coleman, Inc., bkgrd. 299: G. Buttner/Okapia/Photo Researchers, Inc., b.r.; Roy Morsch/The Stock Market, t.r. 300: Liaison Agency, b.r.; Owen Franken/Corbis., b.l. 301: Ken Karp. 302: Phillip Hayson/Photo Researchers, Inc., t.; Richard Hamilton Smith/Corbis., b. 303: Ray Ellis/Photo Researchers, Inc. 304: Will McIntyre/Photo Researchers, Inc. 305: Ken Karp. 306: Bruce Byers/FPG, b.c.; Ken Karp, t.b.l. 306–307: Jeffrey Sylvester/FPG. 307: Norman Owen Tomalin/Bruce Coleman, Inc., r. & b.;Steve Kline/Bruce Coleman, Inc., inset. 308: Lawson Wood/Corbis. 309: Ken Karp. 310: PhotoDisc. 311: Ken Karp. 312: PhotoDisc. b.l Stuart Cahill/AFP/BETTMAN. 315: PhotoDisc. 316: David Sucsy/FPG bkgrd.; Barbara Comnes, inset.

Unit 6: 321: Craig K. Lorenz/Photo Researchers, Inc., bkgrd; Richard Price/FPG, inset. 322: Renee Lynn/Photo Researchers, Inc. 323: Ken Karp. 324: Gary Randall/FPG, b.; Lee Foster/Bruce Coleman, Inc., t. 329: Jon Feingersh/The Stock Market. 330: George F. Mobley, t.; 1998 Comstock, Inc., inset. 331: Emory Kristof. 332: Tim Davis/Photo Researchers, Inc. 333: Ken Karp. 334: Gary Meszaros/Visuals Unlimited. 336: Farrell Grehan/Photo Researchers, Inc., r. ; Rod Planck/Photo Researchers, Inc., l. 337: Ken Karp. 340-341: clockwise from top: Charles Gold/The Stock Market; Denise Cupen/Bruce Coleman, Inc.; Roy Morsch/The Stock Market. Don Mason/The Stock Market; Ed Bock/The Stock Market; Elaine Twichell/Dembinsky Photo Assoc.; J. Barry O'Rourke/The Stock Market; J Sapinsky/The Stock Market; Rex A. Butcher/Bruce Coleman, Inc. 340–341: Telegraph Colour/FPG. 342: Ken Karp. 343: Ken Karp. 345: Dennie Cody/FPG, l; DiMaggio/Kalish/The Stock Market, r. 346: Paul A. Zahl, l; William E. Townsend/Photo Researchers, Inc., r. 347: Arthur Norris/Visuals Unlimited, l.; Biophoto Associates/Photo Researchers, Inc., r. 348: Ken Karp, b.; Zig Leszcynski/Animals Animals, t. 349: Lynwood Chase/Photo Researchers, Inc. 353: Gil Lopez-Espina/Visuals Unlimited, inset; K & K Ammann/Bruce Coleman, Inc., bkgrd. 354: Michael Gadomski/Photo Researchers, Inc. 355: Ken Karp. 356: Joe McDonald/Bruce Coleman, Inc., t.; John Shaw/Bruce Coleman, Inc., b. 358: Ken Karp. 359: Kenneth W. Fink/Bruce Coleman, Inc. 361: Ken Lucas/Visuals Unlimited. 362: Richard Kolar/Animals Animals, l.; Richard & Susan Day/Animals Animals, b. 363: Ken Karp. 365: Barbara Gerlach/Visuals Unlimited, t.; Zefa Germany/The Stock Market, b. 366: A. Cosmos Blank/Photo Researchers, Inc., b.l.; Breck P. Kent/Animals Animals, r.; Robert P. Carr/Bruce Coleman, Inc., b.r. 367: Ken Karp. 369: Emily Stong/Visuals Unlimited. 370: Art Wolfe/Tony Stone Images t.c.; Tom Brakefield/The Stock Market, b. 371: Gerald & Buff Corsi/Visuals Unlimited, t.l.; Stephen Dalton/Photo Researchers, Inc., t.r.; Dan Suzio/Photo Researchers, Inc., b. 372: David Weintraub/Photo Researchers, Inc. 373: Ken Karp. 374: Keith Gunnar/Bruce Coleman, Inc., l.; Phil Degginger/Bruce Coleman, Inc., r. 375: Pat & Tom Leeson/Photo Researchers, Inc. 376: Joe McDonald/Visuals Unlimited. 377: Joe & Carol McDonald/Visuals Unlimited. 378: Ken Karp., b.; Omikron/Photo Researchers, Inc., t. 379: Pat & Tom Leeson/Photo Researchers, Inc. 380: Photo Disc t.; Janis Burger/Bruce Coleman, Inc., t.r.; Jen and Des Bartlett/Bruce Coleman, Inc., b.l.; Tom Van Sant/The Stock Market, bkgrd.

Unit 7: 385: George Schiavone/The Stock Market. 386: Gary Landsman/The Stock Market. 387: Ken Karp. 388: Yoav Levy/Phototake. 389: Barbara Peacock/FPG. 392: Ken Karp. 393: Ken Karp. 394: Michael Townsend/Tony Stone Images, t.; Randy Taylor/Liaison Agency, inset; 395: Bob Daemmrich/The Image Works. 396: David Waldorf/FPG. 397: Ken Karp. 399: David M. Phillips/Visuals Unlimited. 401: Manfred Kage/Peter Arnold, Inc. 402: Ken Karp. 403: Mary Kate Denny/PhotoEdit. 404: CORBIS/BETTMANN–UPI. 405: Sandy Fox/MMSD. 406: Howard Sochurek/The Stock Market, inset; Deborah Gilbert/The Image Bank, b. 407: Telegraph Colour Library/FPG, bkgrd. r. McGraw Hill School Division. 409: Otto Rogge/The Stock Market; t. Tracy/FPG. 410: Joyce Photographics/Photo Researchers, Inc., l.; Steven Needham/Envision, r. 411: Ken Karp. 412–413: Ken Karp. 414: David Young-Wolff/PhotoEdit, t.; Ken Karp, b. 417: Michael Newman/PhotoEdit. 418: NASA/Photri, b.r. & m.r. 419: NASA/Photri, t.r., m.r. & b.r.; NASA/Corbis, inset top. 418–419: Ronald Royer/Photo Researchers, Inc. 420: David Young-Wolff/PhotoEdit. 421: Ken Karp. 422: Michael A. Keller/The Stock Market. 426: Ken Karp. 428: Bkgrd: PhotoDisc.

Handbook: Steven Ogilvy: pp. R6, R8, R12, R14, R15, R26.